THE PHYSICS BOOK

WORKBOOK

UNITS **1** **2**

Scott Adamson
Oliver Alini
Neil Champion
Tara Kuhn

The Physics Book Units 1 & 2 Workbook
1st Edition
Scott Adamson
Oliver Alini
Neil Champion
Tara Kuhn

Contributing authors: Gary Bass, Geoff Cody, Robert Farr, Suzanne Garr,
Roger Walter and Kate Wilson

Publishing editor: Rachel Ford
Project editor: Nadine Anderson-Conklin
Editors: Scott Vandervalk and Kirstie Irwin
Proofreader: Alyssa Lanyon-Owen
Permissions researchers: Catherine Kerstjens and Corrina Gilbert
Cover image: Getty Images/fotojog
Cover design: Chris Starr (MakeWork)
Text design: Watershed Design
Production controller: Karen Young
Typeset by: Macmillan Publishing Solutions

Any URLs contained in this publication were checked for currency during the production process. Note, however, that the publisher cannot vouch for the ongoing currency of URLs.

Acknowledgements

Physics 2019 v1.0 General Senior Syllabus © Queensland Curriculum & Assessment Authority.This syllabus forms part of a new senior assessment and tertiary entrance system in Queensland. Along with other senior syllabuses, it is still being refined in preparation for implementation in schools from 2019.For the most current syllabus versions and curriculum information please refer to the QCAA website https://www.qcaa.qld.edu.au/. The QCAA's permission does not imply permission to reproduce non-QCAA material. For permission to reproduce the material for which the QCAA owns copyright, please contact the QCAA, PO Box 307, Spring Hill Qld 4004; publishing@qcaa.qld.edu.au.

For product information and technology assistance,
in Australia call **1300 790 853**;
in New Zealand call **0800 449 725**

For permission to use material from this text or product, please email **aust.permissions@cengage.com**

ISBN 978 0 17 041255 1

Cengage Learning Australia
Level 7, 80 Dorcas Street
South Melbourne, Victoria Australia 3205

Cengage Learning New Zealand
Unit 4B Rosedale Office Park
331 Rosedale Road, Albany, North Shore 0632, NZ

For learning solutions, visit **cengage.com.au**

Printed in Singapore by 1010 Printing Group Limited.
1 2 3 4 5 6 7 22 21 20 19 18

CONTENTS

UNIT 1 » THERMAL, NUCLEAR AND ELECTRICAL PHYSICS 1

TOPIC 1: HEATING PROCESSES

9780170412551

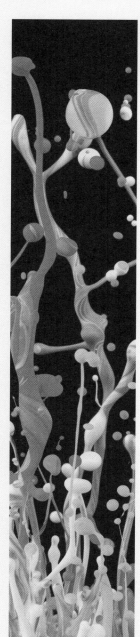

TOPIC 3: ELECTRIC CIRCUITS

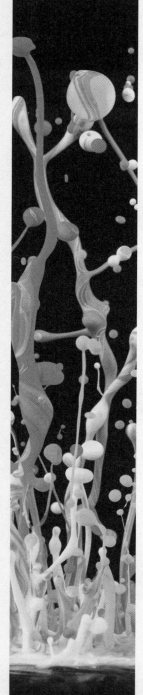

TOPIC 1: LINEAR MOTION AND FORCE

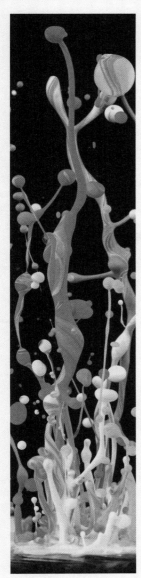

TOPIC 2: WAVES

HOW TO USE THIS BOOK

Learning

The learning section is a summary of the key knowledge and skills. This summary can be used to create mind maps, to write short summaries and as a check list.

Revision

This section is a series of structured activities to help consolidate the knowledge and skills acquired in class.

Evaluation

The evaluation section is in the style of a practice exam to test and evaluate the acquisition of knowledge and skills.

Practice exam

A tear-out exam helps to facilitate preparing and practicing for external exams.

ABOUT THE AUTHORS

Scott Adamson

An experienced maths author, science reviewer, HOD Science and member of the QCAA Physics State Panel and QCAA Science LARG, Scott brings a wealth of knowledge in teaching physics, as well as teaching senior science, to the author team. Scott's dedication to teaching physics has enabled him to lead the *QScience* team in the development of this text.

Oliver Alini

Oliver is a dedicated senior physics teacher and assistant boarding house master. In addition to his experience in teaching physics, Oliver has also been a maths teacher and tutor.

Neil Champion

Neil was directly involved in writing the Australian Senior Physics Curriculum. Neil is an experienced physics author, leading the team on *Nelson Physics for the Australian Curriculum*. Neil has taught physics at both a high-school and a university level, including the physics teaching methodology.

Tara Kuhn

An experienced science teacher, Tara brings her enthusiasm for physics and maths to the writing team.

Some of the material in *The Physics Book Units 1 & 2* has been taken from or adapted from the following: *Nelson Physics for the Australian Curriculum Units 1 & 2* NelsonNet, written by Neil Champion, Geoff Cody, Robert Farr, Suzanne Garr and Roger Walter.

Nelson Physics for the Australian Curriculum Units 3 & 4 NelsonNet, written by Neil Champion, Kate Wilson, Robert Farr, Roger Walter and Gary Bass.

9780170412551

SYLLABUS REFERENCE GRID

UNITS AND TOPICS	THE PHYSICS BOOK UNITS 1 & 2
UNIT 1: THERMAL, NUCLEAR AND ELECTRICAL PHYSICS	
Topic 1: Heating processes	
Kinetic particle model and heat flow	Chapter 1
Temperature and specific heat capacity	Chapter 2
Phase changes and latent heat	Chapter 3
Energy conservation in calorimetry	Chapter 4
Energy in systems – mechanical work and efficiency	Chapter 5
Topic 2: Ionising radiation and nuclear reactions	
Nuclear model and stability	Chapter 6
Spontaneous decay and half-life	Chapter 7
Nuclear energy and mass defect	Chapter 8
Topic 3: Electric circuits	
Current, potential difference and energy flow	Chapter 9
Resistance	Chapter 10
Circuit analysis and design	Chapter 11
UNIT 2: LINEAR MOTION AND WAVES	
Topic 1: Linear motion and force	
Vectors	Chapter 12
Linear motion	Chapter 13
Forces	Chapter 14
Newton's laws of motion	Chapter 15
Topic 2: Waves	
Wave properties	Chapter 16
Sound	Chapter 17
Light	Chapter 18

Physics 2019 v1.0 General Senior Syllabus © Queensland Curriculum & Assessment Authority

THERMAL, NUCLEAR AND ELECTRICAL PHYSICS

- Topic 1: Heating processes

- Topic 2: Ionising radiation and nuclear reactions

- Topic 3: Electric circuits

Kinetic particle model and heat flow

1

LEARNING

Summary

- The kinetic particle model states that all matter is made up of particles that exhibit constant random motion (Brownian motion).
- This motion results in elastic collisions between the particles that transfer energy between them.
- Although energy can be transferred between particles, the total energy of the particles before a collision is equal to the total energy of the particles after the collision.
- The total energy of an individual particle is made up of its kinetic energy and its potential energy.
- The kinetic energy of a particle is the energy it possesses due to its motion.
- The potential energy of a particle is the energy that is stored by moving the particle away from its ideal bond length.
- The temperature of an object is proportional to the total amount of kinetic energy its particles possess.
- The phase (solid, liquid or gas) that a substance is found in is due to the balance between the kinetic energy of its particles and the strength of the intermolecular bonds.
- The internal energy of a substance is the sum of all the kinetic and potential energies of the particles of that substance.
- Heat is the energy that moves from one substance to another because of a difference in temperature between them.
- As a substance in a particular phase is heated, its temperature and therefore its kinetic energy increases.
- If heat is added to a substance that is changing phase, the potential energy of the substance increases; however, the temperature, and therefore the kinetic energy, of the substance does not change.
- The SI unit for energy is the joule (J).
- Heat always moves from an area of high temperature to an area of low temperature.
- Conduction is the transfer of heat through the action of particle collisions.
- Convection is the tendency for hotter particles in a liquid or gas to flow upward while the cooler particles sink downward.
- Radiation is the transmission of heat through electromagnetic radiation.

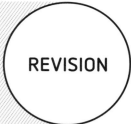
1.1 | Kinetic particle model of matter

The kinetic particle model of matter is used to describe the motion of the particles making up matter. Its main precepts include the fact that particles are in constant motion and can exhibit varying amounts of kinetic energy and potential energy due to their velocities and positions respectively. As the temperature of a substance increases, its velocity, and hence its kinetic energy, will also increase. As the particles become more energetic, they move further away from each other and increase the amount of potential energy stored in their bonds.

QUESTIONS

1 Provide a definition for each of the following terms by using them in a sentence.

a Matter

_____

_____

b Brownian motion

_____

_____

c Atom

_____

_____

d Molecule

_____

_____

e Elastic collision

_____

_____

f Kinetic energy

_____

_____

g Potential energy

_____

_____

2 Draw a 3-circle Venn diagram in the space below. Label the circles: Solid, Liquid and Gas. Match and place the following physical traits of matter with their correct physical trait in the Venn diagram. Note that each state should contain four traits.

Fixed volume

No fixed volume

Fixed shape

No fixed shape

Low kinetic energy

Mid-range kinetic energy

High kinetic energy

Intermolecular bonding

3 If the average kinetic energy of solid water is measured at two different temperatures, a graph can be plotted as shown in Figure 1.1.1 below. Draw a line of best fit for this data and compare the type of relationship shown by this line to your understanding of the relationship that theoretically exists between average kinetic energy and temperature. Identify what the gradient of the line represents in the equation.

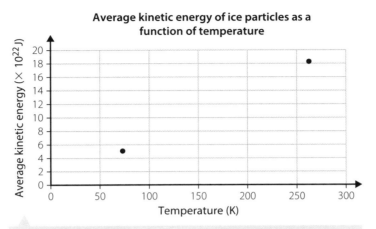

FIGURE 1.1.1 Graph of average kinetic energy versus temperature

_____

_____

_____

_____

_____

_____

_____

_____

_____

_____

1.2 | The energy model

The internal energy of a substance is equal to the sum of the total kinetic energy of its particles due to their motion and the total potential energy stored due to the particles' movement away from their ideal intermolecular bond length. Even though the average kinetic and potential energies of individual particles may vary, the internal energy will remain constant, and the energy will move between the particles through elastic collisions. When energy in the form of heat is added to a substance, the temperature (and therefore the kinetic energy) of the substance will increase until it begins to change phase, at which point the temperature (and therefore the kinetic energy) remains constant while the potential energy increases.

QUESTIONS

1 Select the most appropriate words from the given list in order to fill in the gaps appearing within the following statements.

kinetic	transformed	elastic	heat	intermolecular bonds
transferred	temperature	internal	sum	conserved

   a All forms of energy can be _____ from one form to another and _____ from one place to another.

   b In an _____ collision, the total energy of the colliding particles is always _____.

   c _____ energy is the energy a body possesses due to its motion.

   d The potential energy of a substance is the energy that is stored in the way that the particles are connected to each other through the existence of _____.

   e The _____ energy of a substance is equal to the _____ of the kinetic and potential energies of its particles.

   f The _____ of a substance is a measure of the average kinetic energy of its particles.

   g _____ is energy that spontaneously moves between substances because of a difference in temperature between them.

2 Create a typical heating curve that shows the temperature of a solid to which heat is added until it becomes a gas. Explain the energy changes that are happening to the particles of the substance during each section of the curve.

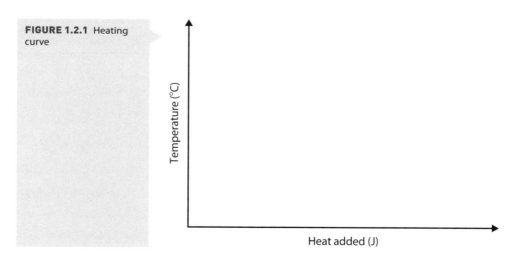

FIGURE 1.2.1 Heating curve

Temperature (°C)

Heat added (J)

1.3 | Heat transfers

Heat always spontaneously moves from areas of higher temperature to areas of lower temperature through the processes of conduction, convection and/or radiation. In conduction, when particles of higher temperature (and kinetic energy) elastically collide with particles of lower temperature, they transfer some of their kinetic energy. Over a period of time, this results in an even distribution of kinetic energy and, therefore, temperature. Convection occurs in fluids and gases when higher-temperature particles flow upwards and lower temperature particles flow downwards, creating convection currents. Radiation is the transfer of energy by the emission of electromagnetic radiation.

QUESTIONS

1   Illustrate each method of heat transfer by drawing a diagram:

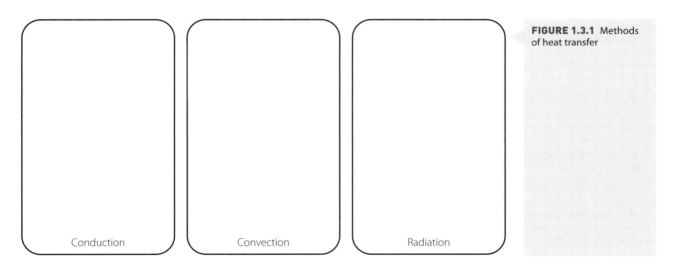

FIGURE 1.3.1 Methods of heat transfer

2 Draw a diagram of a Dewar flask and label the features that help prevent heat transfer out of the flask.

3 To which of the three forms of heat transfer do the following statements relate?

a Holding a cube of ice will cool your hand.

_____

b An oil heater will heat the air of a room.

_____

c A pot placed on a stove will heat up.

_____

d Standing near a camp fire will warm you.

_____

e A sunny day will result in an increase in air temperature.

_____

f Hot, moist air rises high into the atmosphere.

_____

9780170412551

1 Which of the following options is a good description of Brownian motion?

   A The predictable motion of a substance

   B The predictable motion of a single particle in a substance

   C The random motion of a substance

   D The random motion of the particles within a substance

2 In which of the following states of matter is the average kinetic energy of the particles in a substance insufficient to overcome the intermolecular forces holding them in place?

   A Solid

   B Liquid

   C Gas

   D Plasma

3 Which of the following options describing the energy of two particles before and after an interaction is indicative of an elastic collision?

	INITIAL KE OF PARTICLE A	INITIAL KE OF PARTICLE B	FINAL KE OF PARTICLE A	FINAL KE OF PARTICLE B
A	1 J	2 J	1 J	1 J
B	1 J	2 J	2 J	1 J
C	1 J	2 J	3 J	1 J
D	1 J	2 J	4 J	1 J

4 Which of the following regions of the electromagnetic spectrum is the most energetic?

   A Radio waves

   B Visible light

   C X-rays

   D Gamma rays

5 What is the name of energy that is in the process of being transferred from one substance to another because of a temperature difference between them?

   _____

6 Name the property of a substance that is known to increase in direct proportion to the average kinetic energy of its particles.

   _____

7 Explain why it is correct to say that a substance that is not being heated or cooled has a constant internal energy even though the kinetic and potential energies of its individual particles may be changing.

_____

_____

_____

_____

_____

8 If no net temperature change is observed when heat is added to a substance, classify the process the substance must be undergoing and describe what is happening to the energy of the individual particles.

_____

_____

_____

_____

_____

9 Explain why metals are said to be good conductors of heat.

_____

_____

_____

_____

_____

10 When heat from the Sun strikes a body of water, it will often result in precipitation. Explain the heat transfers that must take place for this to happen.

_____

_____

_____

_____

_____

_____

_____

_____

9780170412551

2 Temperature and specific heat capacity

LEARNING

Summary

▶ Qualitative temperature scales give a numerical value to temperature.

▶ The most common temperature scales are the Celsius scale and the Kelvin scale.

▶ $T_K = T_C + 273$ and the difference between each increment on the Celsius scale is the same as the difference between each increment on the Kelvin scale.

▶ The lowest temperature on the Kelvin scale is zero; absolute zero temperature on the Kelvin scale coincides with the temperature at which the Brownian motion of particles stops.

▶ The second law of thermodynamics states that heat always moves from a hotter object to a colder object.

▶ A thermometer is an instrument used to measure the temperature of an object.

▶ There is always some uncertainty in any measurement.

▶ Accuracy is an indication of how close a measurement is to the true value.

▶ Precision is an indication of how reproducible a measurement is.

▶ The uncertainty in a measurement must always be reported, either in the form of an absolute uncertainty or as a relative uncertainty; both give an indication of a range of values that the true value should be between.

▶ All measurements must also always contain a unit. There are seven fundamental SI units and various derived units that can be used.

▶ The addition of heat to a substance results in an increase of the average kinetic energy of its particles, which results in an increase in the temperature of the object.

▶ The amount of heat required to raise the temperature of an object is proportional to its mass, the material it is made from and the size of the temperature change it is undergoing:
$$Q = mc\Delta T$$

▶ The specific heat capacity, c, of a substance gives an indication of how much heat is required to raise the temperature of 1 kg of a substance by 1° Celsius.

▶ Water has a relatively high specific heat capacity, and as a result it takes a significant amount of heat energy to raise its temperature. Such a substance is termed a heat sink.

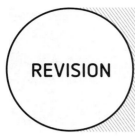
2.1 Converting temperature

Although temperature is ultimately a measurement of the average kinetic energy of the particles in an object, the difficulty in measuring this kinetic energy necessitates a numerical scale by which to measure temperature. The two most common temperature scales are the Celsius and Kelvin scales, both of which use the boiling and freezing point of water as reference points.

QUESTIONS

1 A manufacturer of a thermometer failed to put numerical values on the thermometer (Figure 2.1.1), but included the freezing and boiling points of water. Draw in your own scale indicating the following temperatures: 0°C, 10°C, 20°C, 50°C, 75°C and 100°C.

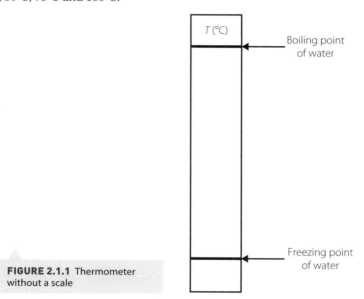

FIGURE 2.1.1 Thermometer without a scale

2 Complete the following statement about absolute zero by inserting the given terms into the spaces provided.

colder	Celsius	hypothesised	second	object	absolute
Kelvin	lowest	ideal		kinetic	transfer	hotter

Lord Kelvin _____ that the volume of an _____ gas should condense to zero at −273° _____ .

The only way that this could occur is if the _____ energy of the particles is reduced to zero. He used this

_____ temperature as the basis for his new temperature scale, the _____ scale. Unfortunately,

the _____ law of thermodynamics states that heat always moves from a _____ object to

a _____ object, which means that we can never reach this _____ of all temperatures, because,

to cool down, the object would need to _____ its energy to something colder than it. So to reach absolute

zero, we would need another _____ that is already at absolute zero, and we are yet to find such an object.

3 Complete the following table showing the relationship between degrees Celsius and Kelvin for some significant temperatures.

PROPERTY	T_C (°C)	T_K (K)
The freezing point of water	0	
The boiling point of water		373
The freezing point of milk	−4	
The average temperature of space		2.7
Average surface temperature of Earth	15	
Average surface temperature of the Sun		5778
Average core temperature of the Sun	1.57×10^7	

2.2 Collecting data

A thermometer measures the change in some temperature-dependent property of a sample to calculate the temperature of an object. The type of thermometer used depends on the type of object, the accuracy and the precision required.

QUESTIONS

1 Match each of the thermometer types appearing in the left column of the table below to the properties that they use to measure temperature.

THERMOMETER TYPE	PROPERTY
Thermostat	Uses the electromagnetic radiation from a surface to measure temperature on the absolute temperature scale
Mercury in glass	Uses different temperature-dependent electrical properties of different metals that are brought into contact
Bimetallic strip	Uses variation in coefficients of expansion between two different metals to detect temperature changes
Thermal paint	Uses variation in electrical resistivity of a material with temperature
Digital	Uses the variation in resistivity of a material with temperature; the greater the resistance the lower the current
Thermocouple	Uses colour change with temperature
Infrared	Uses different coefficients of expansion between mercury and glass

2.3 Practical skills: measurements

A measurement involves comparing a property of interest of an object to some known scale or standard. For the measurement to be reported as validly as possible, the value of the measurement, the units it is using and the uncertainty in the measurement must all be reported. The SI system outlines a set of standard protocols to be used when reporting measurements, to be able to communicate them as reliably as possible.

QUESTIONS

1 Create a glossary containing the following terms (include formulas where appropriate).

TERM	DEFINITION
Measurand	
Measurement result	
True value	
Precision	
Accuracy	
Uncertainty	
Indication values	
Mean value	
Accepted value	
Scientific form	
Random error	
Systematic error	
Absolute uncertainty	
Parallax error	
Confidence interval	
Maximum value	
Minimum value	
Significant figures	
Percentage uncertainty	
Relative uncertainty	
Proportional error	
Percentage error	
Fundamental units	
Derived units	

9780170412551

2 A digital thermometer displaying numerals in the hundredths column is used in an experiment to measure the temperature of the boiling and freezing points of water. The following data is produced:

Freezing point temperature (°C)	0.45
Boiling point temperature (°C)	99.35

Use the data provided to report the following values.

a The absolute uncertainty in both measurements

_____

_____

b The percentage uncertainty in both measurements

_____

_____

_____

_____

c The percentage error in the boiling point measurement if the accepted value is 100°C

_____

_____

_____

_____

2.4 | Changes in temperature

When heat is added to the surface of a substance, the particles in contact with the heat source will gain kinetic energy and there is an immediate increase in the average kinetic energy of the particles of the substance. These higher-energy particles will then go on to elastically collide with other particles in the substance and so transfer the kinetic energy through the substance while more kinetic energy is gained at the surface. In this way, the kinetic energy, and therefore the temperature, of the whole substance increases. The rate of this process is dependent on the thermal conductivity of the substance.

QUESTIONS

1   On Figure 2.4.1 below, use arrows to show what is happening, in terms of heat flow, kinetic energy and temperature, to an object that is placed in contact with a heat source.

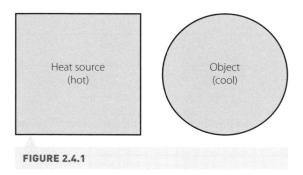

FIGURE 2.4.1

2   On Figure 2.4.2 below, use arrows to show what is happening, in terms of heat flow, kinetic energy and temperature, to a warm object (B) which has been placed in contact with a cool object (A).

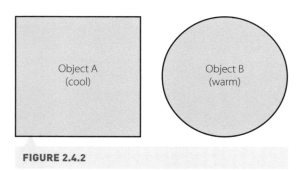

FIGURE 2.4.2

9780170412551

2.5 | Specific heat capacity and proportionality

The amount of heat required to raise the temperature of a substance, or the amount of heat released when a substance is cooled, is proportional to the mass of the substance, the make-up of the substance and the degree of temperature change. This is shown in the formula $Q = mc\Delta T$. The specific heat capacity (c) is a measure of the amount of energy required to increase the temperature of a 1 kg sample of a substance by 1°C (or 1 K). It is a material-specific constant that is dependent on the physical properties of the substance. Liquid water has a relatively high specific heat capacity and therefore requires a significant amount of heat energy to increase its temperature. Such a substance is referred to as a heat sink.

QUESTIONS

1 Place the following substances in order of their increasing potential use as a heat sink.

SUBSTANCE	SPECIFIC HEAT CAPACITY (J kg^{-1} K^{-1})	ORDER (1−6)
Steam	2000	
Cooking oil	2800	
Soil	800	
Ice	2100	
Liquid water	4180	
Air	1000	

2 The following data shows the amount of heat added to various masses of a substance in order to increase its temperature by 1°C.

MASS OF SUBSTANCE (kg)	HEAT ADDED (J)
0.05	40
0.15	120
0.5	400
1	800
1.25	1000
2	1600

a Use the provided data to create a graph showing the relationship between mass and heat added.

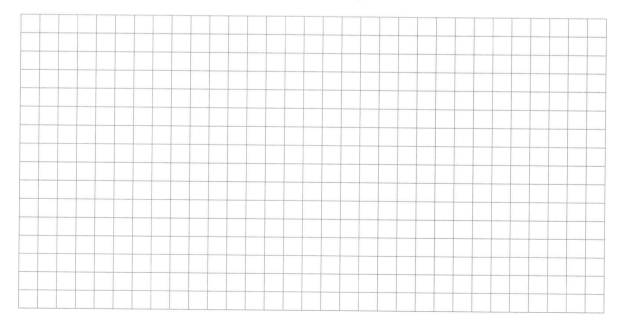

b Explain how the data supports the statement that the amount of heat required to raise the temperature of a substance is proportional to the mass of the substance.

_____

_____

_____

c Determine the specific heat of the substance.

_____

_____

_____

d Suggest the type of substance used.

_____

2.6 | Solving problems: specific heat capacity

The specific heat formula, $Q = mc\Delta T$, can be used to solve many problems that involve a change of temperature of a substance with the addition or removal of heat. In solving these problems, it is particularly important to pay close attention to the units used for each term, to ensure that they agree with each other.

WORKED EXAMPLE

Calculate the specific heat capacity of a 550 g substance if it requires 25 410 J of heat to raise its temperature from 33°C to 75°C.

ANSWER

1 Use the specific heat equation:

$$Q = mc\Delta T$$

2 Rearrange for the unknown:

$$c = \frac{Q}{m\Delta T}$$

3 Insert known values:

$$c = \frac{25\,410 \text{ J}}{0.55 \text{ kg} \times (75° - 33°)}$$

4 Give the answer with the correct units and the correct number of significant figures:

$c = 1100 \text{ J kg}^{-1}°\text{C}^{-1}$

QUESTIONS

1 Determine the specific heat capacity of a substance in $J\,kg^{-1}\,K^{-1}$ if its specific heat capacity is known to be $0.67\,kcal\,kg^{-1}\,K^{-1}$.

_____

_____

_____

_____

_____

_____

2 If 1.2 kg of ice is heated from $-10.0°C$ to $0.0°C$, determine the amount of heat that must have been added.

_____

_____

_____

_____

_____

_____

3 Determine the amount of heat released when 0.600 kg of steam is cooled from $116.0°C$ to $100.0°C$.

_____

_____

_____

_____

_____

_____

4 Determine the mass of water in a sample if its temperature increased by 6.7°C on the addition of 187 kJ of heat.

_____

_____

_____

_____

_____

_____

_____

5 Determine the specific heat capacity of a 6.5 kg sample of an unknown substance if 16.4 kcal of added heat increases its temperature by 17°C.

_____

_____

_____

_____

_____

_____

9780170412551

1 Which of the following temperatures on the Kelvin scale corresponds to the freezing point of water?

   A 0K

   B 100K

   C 273K

   D 373K

2 Which of the following numbers contains four significant digits?

   A 1.02

   B 0.102

   C 1.020

   D 10.2×10^{-1}

3 Which type of uncertainty is the uncertainty in the measurement 6.3±0.5% an example of?

   A Absolute uncertainty

   B Relative uncertainty

   C Percentage uncertainty

   D Percentage error

4 Which of the following units for the specific heat capacity of a substance consists entirely of fundamental units according to the SI?

   A $Jkg^{-1}K^{-1}$

   B $Jg^{-1}K^{-1}$

   C $Jkg^{-1}{}^{\circ}C^{-1}$

   D $kJkg^{-1}{}^{\circ}C^{-1}$

5 Which of the following substances would be most useful as a heat sink?

   A Ice

   B Soil

   C Crown glass

   D Lead

6 What term is given to data that is descriptive rather than numerical?

_____

7 What type of error would result in all measured temperatures in an experiment being 3.2°C higher than their actual value?

_____

8 If a certain mass of a substance undergoes an increase in temperature when heat is added, by how much will the temperature of the same substance increase if there is twice as much matter and the same amount of heat is added?

_____

_____

9 Suggest the reasons for why scientists prefer to use the Kelvin scale when describing the temperature of a substance.

_____

_____

_____

_____

_____

10 During the hardening process of concrete, a chemical reaction takes place that heats the wet concrete. Once the reaction is complete, the temperature returns to air temperature and the concrete hardens dry. Given this information, explain why it is important to include expansion gaps when pouring concrete.

_____

_____

_____

_____

_____

11 Convert $380\,\mathrm{J\,kg^{-1}\,K^{-1}}$, the specific heat capacity of copper, into $\mathrm{kcal\,kg^{-1}\,K^{-1}}$.

_____

_____

_____

_____

_____

12 Deduce the amount of energy that would be required to heat a 650 g sample of copper at 31°C to 125°C.

_____

_____

_____

_____

13 Calculate the specific heat capacity of a sample if 38 000 J of heat is released when 2.1 kg of the sample cools by 24.5°C.

_____

_____

_____

_____

_____

14 A current of 2.0 A passes through a heating element as a 12 V power source is turned on. Calculate the final temperature of a 350 mL sample of pure water initially at 15°C, after 12 minutes of heating.

_____

_____

_____

_____

_____

15 A 150.0 g sample of lead at 105°C is added to a 150.0 g aluminium calorimeter containing 180.0 mL of water at 15.0°C. Calculate the final temperature of the metal if the sample and the calorimeter are allowed to stand until their temperatures equilibrate.

_____

_____

_____

_____

_____

16 Explain, in terms of energy and collisions, what happens to the particles of a substance when it undergoes heating.

_____

_____

_____

_____

_____

3 Phase changes and latent heat

LEARNING

Summary

- A phase change is a change in the physical state (solid, liquid, gas) of a substance.
- A solid undergoes melting to become a liquid and a liquid undergoes solidification into a solid.
- A liquid undergoes vaporisation to become a gas and a gas undergoes condensation to become a liquid.
- A solid will undergo sublimation to become a gas and a gas will undergo deposition to become a solid.
- Evaporation is the gradual process by which individual high-energy particles in a liquid escape to become gaseous at a temperature lower than the boiling point of the liquid.
- A heating curve is a graph of heat added vs temperature of a substance; phase changes are indicated by horizontal lines.
- During a phase change, the temperature of a substance does not increase.
- During a phase change, the added heat causes the breaking of intermolecular bonds; once all of these bonds are broken the phase change is complete.
- Latent heat is the amount of heat that must be added to a substance to cause a phase change.
- The specific latent heat is the energy per kilogram of a substance required to cause a phase change.
- The specific latent heat of vaporisation is the heat required for 1 kg of a liquid to become a gas; it is also equal to the heat released by 1 kg of a gas when it becomes a liquid.
- The specific latent heat of fusion is the heat required for 1 kg of a solid to become a liquid; it is equal to the heat released when 1 kg of a liquid becomes a solid.
- $Q = mL$ is the specific heat equation showing that the heat required (Q) for a phase change is equal to the latent heat of that phase change (L) times the mass of the substance.

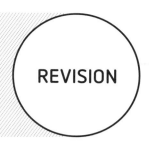

3.1 The process of state change

During a phase change, the physical state of a substance will switch between solid, liquid and gas, with the name of the state change (solidification, melting, vaporisation, condensation, sublimation or deposition) indicating between which two phases the substance is moving. During a phase change, the particles of a substance become energetic enough to break free from their intermolecular bonds. These particles then collide with other particles in a process that continues until all bonds have been broken. During this time, the average kinetic energy of the particles, and therefore the temperature, does not increase; rather, the added heat energy is stored in an increase of potential energy.

QUESTIONS

1   On the diagram in Figure 3.1.1, name the phase changes occurring.

2   Label the boxes in the flow diagram (Figure 3.1.2) with the letters **A** to **F** indicating what occurs to a substance on a particle level during a phase change.

   A   Individual particles gain enough kinetic energy to overcome their intermolecular bonds.

   B   Phase change is complete.

   C   This process continues until all particles have enough kinetic energy to overcome their intermolecular bonds.

   D   Heat is added to a system.

   E   Phase change begins.

   F   Individual energetic particles collide with other less energetic particles, transferring energy through elastic collisions.

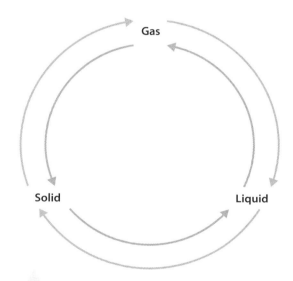

FIGURE 3.1.1 Phase change cycles

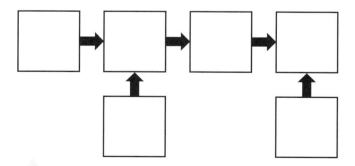

FIGURE 3.1.2 Flow chart showing particle interactions during a phase change

3 Create a heating curve for a solid that goes through the various phase changes to a gas and label the different sections of the curve with the following terms.

A Boiling

B Heating of the solid

C Heating of the gas

D Melting

E Heating of the liquid

9780170412551

3.2 | Defining specific latent heat

The heat required to enact a phase change is dependent upon the type of substance, the mass of the substance and the phase change that the substance is undergoing. This is summarised by the formula $Q = mL$, where Q is the heat added or released during a phase change, m is the mass of the substance, and L is the specific latent heat of the substance and phase change. The specific latent heat can be further categorised as the latent heat of fusion for changes between the solid and liquid phases, and the specific latent heat of vaporisation for changes between the liquid and gas phases.

QUESTIONS

1   A student investigation was carried out by applying heat at a steady rate of $500\,\mathrm{J\,s^{-1}}$ to different masses of ice at 0°C, which had been placed in an insulated container to prevent heat loss to the surroundings. This heat was applied to the sample until the point at which all of the ice was observed to have melted. The mass of the ice and the melting time were recorded in Table 3.2.1.

TABLE 3.2.1

MASS OF ICE AT 0°C (kg)	TIME FOR MELTING (s)	TOTAL ENERGY INPUT (kJ)
0.10	66	
0.22	144	
0.39	176	
0.52	350	
0.64	434	

a   Complete the final column of Table 3.2.1.

b   Graph the data on the grid provided below.

c Are there any anomalies in the data? Hypothesise a reason for their existence.

_____

_____

_____

_____

_____

d Analyse the graph and insert a line of best fit.

e Calculate the equation of the line of best fit.

_____

_____

_____

f Knowing that $Q = mL$, to what property of ice does the gradient refer?

_____

_____

g Compare your result from part **e** to the accepted value of L for ice.

_____

_____

3.3 Solving problems: specific latent heat

The specific heat formula $Q = mL$ can be used to solve many theoretical and real-life problems involving phase changes. The value of L for a particular substance undergoing a specific phase change can be found on tables or may be given to you in a question. Care must be taken to ensure that all units are in standard SI form.

WORKED EXAMPLE

Determine the amount of heat that must be added to 650 g of ice at its melting point to completely liquefy it to water.

ANSWER

1 Apply the latent heat of fusion formula:

$Q = mL_f$

2 Insert known values:

$Q = 0.65 \, \text{kg} \times 334\,000 \, \text{J kg}^{-1}$

3 Calculate the answer:

$Q = 217\,100 \, \text{J}$

4 Give the answer with the correct units and the correct number of significant digits:

$Q = 220 \, \text{kJ}$

9780170412551

QUESTIONS

1 Determine the amount of heat that must be added to 2.3 kg of aluminium, which is at its melting point, for it to completely liquefy.

_____

_____

_____

_____

_____

_____

2 If 217 100 J of heat is released by a sample of steam when it condenses to water, determine the mass of the sample.

_____

_____

_____

_____

_____

_____

_____

3 Gold has a latent heat of fusion of $16000\,\text{cal}\,\text{kg}^{-1}$. Determine the heat required to liquefy a 200 g gold ring completely.

_____

_____

_____

_____

_____

_____

_____

3.4 | Phase changes

The latent heat of the phase change of a substance can be determined by adding heat to a sample before, during and after the transition temperature. If a constant heat source is applied to an insulated substance, a heat curve showing how the temperature of the sample alters during the addition of heat can be constructed. This can then be analysed to determine the latent heat for the sample undergoing the phase change.

QUESTIONS

1 The temperature of an insulated 150g sample of ice initially at −4.0°C is taken at 1 minute intervals, when a 60J s^{-1} immersion heater is placed into it, and the data is recorded in Table 3.4.1.

TABLE 3.4.1 Measuring the latent heat of a phase change

TIME (s)	TEMPERATURE (°C)	HEAT ADDED (kJ)
0	−10	
60	−1.7	
120	0	
180	0	
240	0	
300	0	
360	0	
420	0	
480	0	
540	0	
600	0	
660	0	
720	0	
780	0	
840	0	
900	0	
960	0	
1020	0	
1080	0	
1140	0	
1200	1	
1260	5.3	
1320	9.6	
1380	10.1	
1440	18.1	

9780170412551

a Complete Table 3.4.1 by indicating the amount of heat added at each time.

b Create a heating curve for the data on the graph paper below.

c Identify any anomalies in the data and suggest a reason for their existence.

_____

_____

_____

_____

_____

d Calculate the latent heat of fusion of the sample and compare this to the accepted value.

_____

_____

_____

_____

_____

e Evaluate the experiment and suggest any improvements that can be made in future tests.

_____

_____

_____

_____

_____

1   Which of the following is a phase change associated with a solid?

   A   Condensation

   B   Boiling

   C   Sublimation

   D   Evaporation

2   Heat energy added to a substance that is undergoing a phase change results in an increase in which of the following properties of the particles of that substance?

   A   Kinetic energy

   B   Potential energy

   C   Velocity

   D   Temperature

3   Which of the following solids would require the greatest amount of energy to cause it to completely liquefy?

   A   1 kg of copper

   B   1 kg of lead

   C   1 kg of ice

   D   2 kg of ice

4   What is the name given to the process by which a gaseous substance undergoes a phase change directly to a solid?

_____

5   Would you use the specific latent heat of fusion or the specific latent heat of vaporisation to calculate the amount of energy released when a sample of liquid lead solidifies?

_____

6   Describe the differences between evaporation and vaporisation.

_____

_____

_____

_____

_____

_____

_____

7 Draw a typical heating curve for a sample of water that starts as a solid and ends as a gas. Identify the phases and processes that each region of the curve represents.

8 Determine the amount of energy that must be added to a 2.3 kg sample of solid copper if it is to completely liquefy.

9 Determine the amount of heat released when 650 g of liquid cobalt solidifies, given that it has a specific latent heat of fusion of $58 \, \text{J g}^{-1}$.

10 Determine the latent heat of fusion of ether if 307 kJ of heat is required to melt a 3.2 kg sample.

11 Determine whether or not the amount of heat released by the condensation of 200 g of steam into water would be sufficient to melt a 1.0 kg sample of aluminium.

_____

_____

_____

_____

_____

_____

_____

_____

_____

9780170412551

LEARNING

Summary

▶ When two or more objects in thermal contact are at the same temperature, they are said to be in thermal equilibrium.

▶ At thermal equilibrium, the net heat transferred between objects is equal to zero.

▶ When two objects of different temperatures are placed in thermal contact, the kinetic energy of the object with the higher temperature will be transferred through collisions between particles to the object at the lower temperature.

▶ At thermal equilibrium, the temperatures of the objects are the same, along with the average kinetic energy of the particles.

▶ The zeroth law of thermodynamics states that if two objects are in thermal equilibrium with a third object then they must be in thermal equilibrium with each other.

▶ A system is an object or set of objects that is under investigation.

▶ An open system is one that can transfer mass and energy to its surroundings.

▶ A closed system can transfer energy, but not mass, to its surroundings.

▶ An isolated system can transfer neither mass nor energy to its surroundings.

▶ If two objects form an isolated system, the heat lost by one object must exactly equal the heat gained by the other.

▶ A calorimeter is an experimental device that forms an isolated system and can be used to calculate the specific heat of a substance.

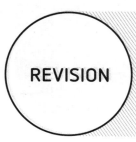
4.1 Thermal equilibrium and the energy of particles

When two objects of different temperature (and therefore average kinetic energy) are put into thermal contact, heat is transferred from the hotter object to the cooler one until both reach the same temperature. They are then said to be in thermal equilibrium. At this point, even though heat is still being transferred between them, the net heat flow between them is zero, and so they remain at the same temperature.

QUESTIONS

1 Complete the following sentences.

Two objects that are at the same temperature are said to be in _____ equilibrium.

A hotter object will transfer _____ energy to a cooler object through elastic _____ .

The _____ lost by the hotter object is equal to the heat gained by the cooler object.

At thermal equilibrium, the net heat flow between the objects is equal to _____, and the average kinetic energy of the two objects is _____ .

2 Draw a diagram showing the flow of heat between an ice cube and a cup of hot water and explain the process.

_____

_____

_____

_____

3 Show the heat flow occurring between a teaspoon and a cup of tea that are at thermal equilibrium. Explain the process.

_____

_____

_____

_____

_____

4.2 | Achieving thermal equilibrium

When two objects of different temperature are placed in thermal contact with each other, particles at the surface of the hotter object, which have a higher average kinetic energy than the particles in the colder object, collide elastically with the particles in the colder object and as a result transfer some of their kinetic energy to the colder object. This results in the average kinetic energy (and the temperature) of the colder object increasing and the average kinetic energy of the hotter object decreasing. This process continues until the two objects have equal average kinetic energies and are therefore in thermal equilibrium.

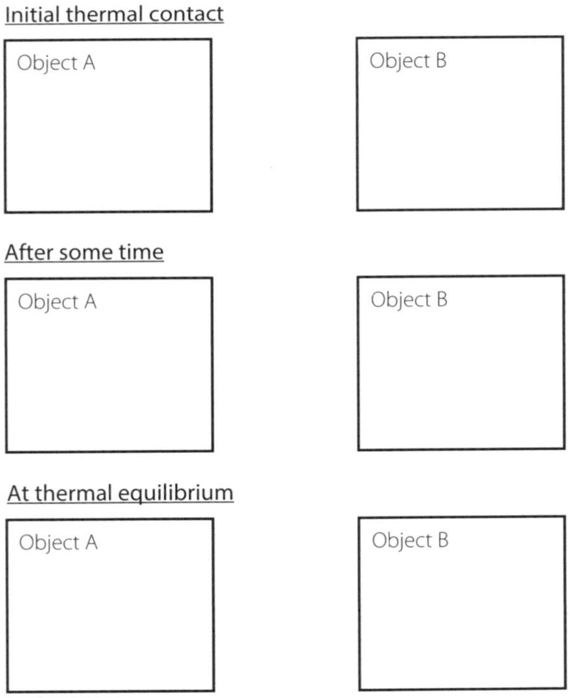

Initial thermal contact

| Object A | Object B |

After some time

| Object A | Object B |

At thermal equilibrium

| Object A | Object B |

FIGURE 4.2.1 Heat flow between objects of different temperatures

QUESTIONS

1 Complete the flow diagram representing the heat flow between objects of different temperatures during the process in which they reach thermal equilibrium (Figure 4.2.1). Assume object A is initially hotter than object B. Show the relative sizes of the heat flow between the two objects and the overall net direction of heat flow between them. Provide a statement about the temperature of each object and a statement about the average kinetic energy of the particles of each object.

2 Create a sentence that summarises the zeroth law of thermodynamics and explains its usefulness, using the following words:

zeroth law of thermodynamics	thermal equilibrium	object
everyday experience	temperature	thermal contact

_____

_____

_____

_____

_____

4.3 | Solving problems: thermal equilibrium and the spontaneous transfer of heat

A system is any object or set of objects that is being investigated. An open system can transfer energy and mass in and out of it, a closed system can transfer energy but not mass in or out of it, and an isolated system can transfer neither energy nor mass into or out of it. If an isolated system is being studied, the law of conservation of energy can be applied, and the heat entering one part of a system must be equal to the same amount of heat leaving another part. A calorimeter is an experimental tool that thermodynamically isolates its contents from the outside environment.

WORKED EXAMPLE

If 650 g of water at 10.0°C is added to 1200 g of water at 60.0°C in an isolated system, determine the final temperature reached by the system.

ANSWER

1 Apply the conservation of energy formula since the system is isolated:

$Q_{lost} = -Q_{gained}$

2 Use the specific heat capacity formula:

$m_{hot} c \Delta T_{hot} = -m_{cold} c \Delta T_{cold}$

3 Expand the brackets:

$m_{hot} c(T_{final} - T_{hot}) = -m_{cold} c(T_{final} - T_{cold})$

4 Expand the brackets:

$m_{hot} c T_{final} - m_{hot} c T_{hot} = -(m_{cold} c T_{final} - m_{cold} c T_{cold})$

5 Gather like terms:

$m_{hot} c T_{final} + m_{cold} c T_{final} = m_{hot} c T_{hot} + m_{cold} c T_{cold}$

6 Factorise T_{final} out of the left-hand side:

$(m_{hot} c + m_{cold} c) T_{final} = m_{hot} c T_{hot} + m_{cold} c T_{cold}$

7 Isolate T_{final} on the left-hand side:

$T_{final} = \dfrac{m_{hot} c T_{hot} + m_{cold} c T_{cold}}{(m_{hot} c + m_{cold} c)}$

9780170412551

8 Insert known values:

$$T_{\text{final}} = \frac{1.2 \text{ kg} \times 4200 \ \frac{\text{J}}{\text{kg}^\circ\text{C}} \times 60^\circ\text{C} + 0.65 \text{ kg} \times 4200 \ \frac{\text{J}}{\text{kg}^\circ\text{C}} \times 10^\circ\text{C}}{\left(1.2 \text{ kg} \times 4200 \ \frac{\text{J}}{\text{kg}^\circ\text{C}} + 0.65 \text{ kg} \times 4200 \ \frac{\text{J}}{\text{kg}^\circ\text{C}} \right)}$$

9 Calculate answer:

$T_{\text{final}} = 42.4324^\circ\text{C}$

10 Give answer with correct units and significant figures:

$T_{\text{final}} = 42.4^\circ\text{C}$

QUESTIONS

1 Draw diagrams of each of the three systems in the boxes provided (Figure 4.3.1). Include the possible transfers of energy and mass that can occur for each system.

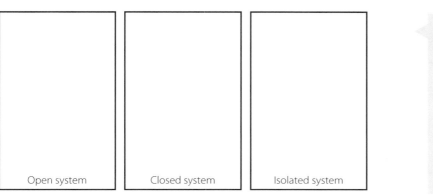

Open system   Closed system   Isolated system

FIGURE 4.3.1 Examples of open, closed and isolated systems

2 Classify each of the following examples of systems as open, closed or isolated.

a Hot water in a thermos flask

_____

b Boiling water in a saucepan with a lid

_____

c A car engine

_____

d A hydraulic fluid that is being compressed

_____

e The universe as a whole

_____

f An air conditioner

_____

3 If 220 g of water at 5.0°C is added to 2240 g of water at 35°C in an isolated system, determine the final temperature reached by the system.

_____

_____

_____

_____

_____

_____

_____

4 A 1.3 kg aluminium frying pan that is at 250°C is to be cooled to 120°C. Determine the amount of ice at −4°C that must be added to it.

_____

_____

_____

_____

_____

_____

_____

5 A 2.0×10^2 g sample of a substance that is initially at 140°C is added to a 150 g sample of water at 30.0°C in a 120 g aluminium calorimeter. It is observed that the system reaches thermal equilibrium at 33.5°C. Determine the specific heat of the unknown substance.

_____

_____

_____

_____

_____

_____

_____

9780170412551

1 Which of the following statements is a correct description of two objects of different temperature placed in thermal contact?

A The net heat flow between them is zero.

B The average kinetic energy of the particles of the two objects is different.

C Heat is flowing from the colder object to the hotter object.

D The potential energies of the two objects are equal.

2 Which of the following is an example of open system?

A A saucepan of water with its lid off that has been placed on a stove

B A saucepan of water with its lid on that has been placed on a stove

C A calorimeter full of water containing a heating element

D A calorimeter full of water that does not contain a heating element

3 If two objects are in thermal equilibrium with a third, which of the following statements is not correct?

A All three objects have equal average kinetic energies.

B The initial two objects are in thermal equilibrium with each other.

C If the three objects are placed in thermal contact, there will be a flow of heat between them.

D If all three objects are placed in thermal contact, their average kinetic energies will remain constant.

4 If two objects are in thermal equilibrium, state the thermodynamic properties of the two objects that must be equal.

_____

5 If a hotter object is placed in thermal equilibrium with a colder object, what will happen to the average kinetic energy of the particles of the colder object?

_____

6 What is the name of a system that can transfer energy and mass to its surroundings?

_____

7 What property of objects does the zeroth law of thermodynamics give a useful description of?

_____

8 Describe the process by which two objects of different temperature reach thermal equilibrium when they are placed in thermal contact with each other.

_____

_____

_____

_____

_____

_____

9 Explain why even though two objects may be in thermal equilibrium there is still heat travelling between them.

_____

_____

_____

_____

_____

_____

10 Explain why if equal amounts of heat are added to two objects initially at the same temperature but not in thermal contact, they will no longer necessarily be in thermal equilibrium.

_____

_____

_____

_____

11 Determine the final temperature of the resulting mixture that occurs when 150g of water at 22°C is added to 350g of water at 62°C (assume no heat is lost to the surroundings).

_____

_____

_____

_____

_____

_____

_____

_____

12 Determine the final temperature of the resulting mixture when 55g of ice at −4°C is added to 250g of water at 85°C (assume no heat is lost to the surroundings).

_____

_____

_____

_____

_____

_____

_____

_____

13 Determine the amount of ice at −4.0°C that must be added to bring an isolated 250g sample of water initially at 65°C to a final temperature of 25°C.

_____

_____

_____

_____

_____

_____

_____

_____

14 A 150 g sample of an unknown substance has been sitting in a water bath at 95°C. If it brings a 250 mL sample of water in a 110 g aluminium calorimeter initially at 25°C to a final temperature of 43°C, determine its specific heat.

_____

_____

_____

_____

_____

_____

_____

9780170412551

5 Energy in systems – mechanical work and efficiency

LEARNING

Summary

▶ Energy can be transferred to another object by the transfer of heat (Q), or the action of work (W).

▶ Work is defined as the action of a force over some distance. $W = F \times s$, where F is the applied force and s is the distance over which the force is applied. Work has units of joules (J) or newton metres (N.m).

▶ Power (P) is defined as the rate at which energy (E) or work (W) is transferred per unit time (t), $P = \dfrac{E}{t} = \dfrac{W}{t}$, and has units of watts.

▶ An external combustion engine produces work through the expansion of a fluid that is heated by the combustion of an external fuel source.

▶ The internal energy (U) of a system is the sum of all kinetic and potential energy present within the system.

▶ The work–energy principle states that if heat (Q) is added or lost by the system, or work (W) is performed by or on the system, the internal energy of the system (U) must change: $\Delta U = Q - W$.

▶ Usable energy is the energy present in a system, which can be used to perform some desired task.

▶ A heat-exchange system is any system that transfers heat from a warmer to a cooler location.

▶ A heat-conversion system transforms the internal energy of a system.

▶ A heat pump is a system that moves heat from one place to another.

▶ All real systems lose energy to their surroundings.

▶ The efficiency (η) of a system is the fraction of the input energy that produces a useful output: $\eta = \dfrac{\text{energy output}}{\text{energy input}} \times \dfrac{100\%}{1}$.

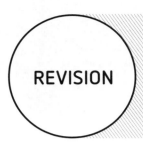

5.1 The capacity to do work

Energy can either enter or leave a system by a transfer of heat (Q) or the action of work (W). Work is defined as the action of a force (F) over some distance (s): $W = F \times s$. The rate at which energy is transferred is called the power (P) of a process, $P = \dfrac{E}{t} = \dfrac{W}{t}$, and has the units of watts (W) or joules per second, Js^{-1}. An external combustion engine, which is an example of a useful system, burns fuel in an external chamber, to provide heat to a fluid in the engine. The fluid then expands and performs work on the pistons of an engine.

QUESTIONS

1 Complete the following statement concerning work and heat by placing the appropriate term in the spaces provided.

In a _____ system, _____ can be added or removed by either _____ or removing _____ or by the system doing _____ or having work done on it. An external _____ engine is an example of such a closed system, whereby _____ is added to the system by the _____ of a fuel in an external combustion _____, which then heats a _____ that expands and does _____ to move a _____. This motion can be used to do work on something that is _____ to the system.

adding	closed	combustion	fluid	burning	work	external
energy	work	external heat	piston	chamber	heat	

2 Draw the flow of heat and work in and out of the diagram of an external combustion engine (Figure 5.1.1).

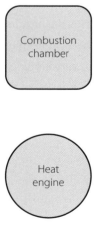

FIGURE 5.1.1 Heat and work flow diagram of an external combustion engine

5.2 | Change in internal energy

If the internal energy of a system is defined as the sum of all the kinetic and potential energy in a system, the work–energy principle can be stated as the change in the internal energy of a system, which is equal to the energy added to the system in the form of heat minus the work done by the system on its surroundings; $\Delta U = Q - W$. Q is positive when heat is added and negative when heat is removed. W is positive when work is done by the system and negative when work is done on the system. In reality, all heat engines are less than 100% efficient due to the loss of heat to the surrounding environment.

QUESTIONS

1 Insert arrows in Figure 5.2.1 to create a flow chart indicating the heat flow in and out of an external combustion heat engine connected to carriages that runs at less than 100% efficiency due to the loss of energy to the surroundings.

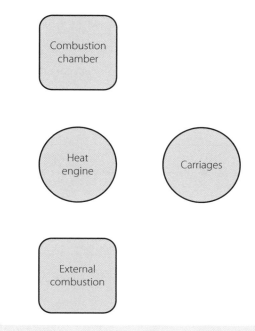

FIGURE 5.2.1 Heat flow diagram for an external combustion engine

2 Determine the change in the internal energy of a system whose output work is 4500 J when 8100 J of heat is added to it.

_____

_____

_____

_____

3 When 125 kJ of heat is added to a steam engine that is at its stable operating temperature, 65 000 J of work is done by the engine. Determine the amount of energy that is being lost from the system in the form of heat released to its immediate environment.

_____

_____

_____

_____

5.3 | Heat loss and usable energy

The production of usable energy in the form of work is the primary goal of any heat engine. Heat engines all operate by converting heat from a fuel source into useful work. In reality, no engine is 100% efficient, and will lose some energy to the environment in the form of heat. A _heat-exchange system_ transfers heat from a warmer to a cooler location. A _heat-conversion system_ transforms the internal energy in a system.

QUESTIONS

1 Write a summary that describes how a reverse-cycle air conditioner manages to heat a house during winter by including the following terms:

> evaporator coil   outside temperature   compressor   internal air temperature   condenser

_____

_____

_____

_____

_____

_____

_____

2 Complete the schematic diagram of the heat pump of a refrigerator (Figure 5.3.1) to show the direction of heat flow and applied work.

FIGURE 5.3.1 Schematic diagram of the heat pump of a refrigerator

Outside of refrigerator (High _T_)

Heat pump

Inside of refrigerator (Low _T_)

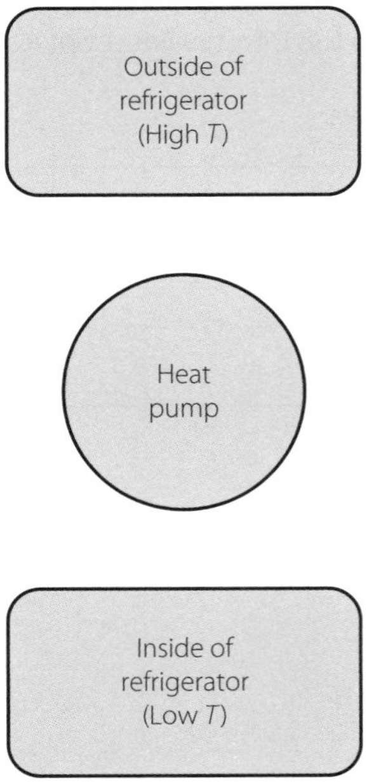

3 Place the following steps (A–D) in a four-stroke internal combustion cycle in their order of operation on the diagram in Figure 5.3.2.

A The combustion stroke – the compressed fuel–air mixture is ignited, causing an explosion that forces the piston downwards.

B The intake stroke – the piston, which is initially is the upward position, moves downwards, and this draws fuel and air into the combustion chamber through the intake valves.

C The exhaust stroke – the piston moves upwards, forcing the exhaust gasses out through the exhaust valves that connect to the exhaust pipe.

D The compression stroke – the piston moves upwards, reducing the volume of the combustion chamber and compresses the air–fuel mixture.

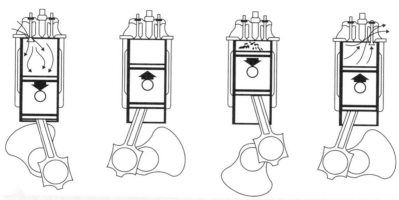

FIGURE 5.3.2 Four-stroke internal combustion cycle

5.4 Efficiency

The energy efficiency, η, is the fraction of the input energy that is being converted to useful output energy:
$\eta = \dfrac{\text{energy output}}{\text{energy input}} \times \dfrac{100\%}{1}$. In the case of heat engines, the input energy is equal to Q_{in} and the output energy is W. If an engine is at its stable operating temperature, $W = Q_{in} - Q_{out}$, efficiency can be calculated as

$$\eta = \left(1 - \frac{Q_{out}}{Q_{in}}\right) \times \frac{100\%}{1}.$$

WORKED EXAMPLE

A heat engine operating at its stable operating temperature has an efficiency of 31% and uses 210kJ of heat energy every second. Determine:

a The amount of work that is output by the system every second

b The amount of heat that is lost to its surroundings in 1.0 minute.

ANSWER

a **1** Use the efficiency formula:

$$\eta = \frac{\text{energy output}}{\text{energy input}} \times \frac{100\%}{1}$$

2 Make the unknown the subject:

$$\text{energy output} = \frac{\eta}{100\%} \times \text{energy input}$$

3 Insert known values:

$$\text{energy output} = \frac{31\%}{100\%} \times 210 \text{ kJ s}^{-1}$$

4 Calculate the answer:

$$\text{energy output} = 65.1 \text{ kJ s}^{-1}$$

5 Answer to correct number of significant figures:

$$\text{energy output} = 65 \text{ kJ}$$

b **1** Use the correct efficiency formula:

$$\eta = \left(1 - \frac{Q_{\text{out}}}{Q_{\text{in}}}\right) \times \frac{100\%}{1}$$

2 Rearrange for the unknown:

$$Q_{\text{out}} = Q_{\text{in}}\left(1 - \frac{\eta}{100\%}\right)$$

3 Insert known values:

$$Q_{\text{out}} = 210 \text{ kJ s}^{-1} \times \left(1 - \frac{31\%}{100\%}\right)$$

4 Calculate the answer:

$$Q_{\text{out}} = 144.9 \text{ kJ s}^{-1}$$

5 Multiply the answer by 60s to get energy lost every minute:

$$Q_{\text{out}} = 144.9 \text{ kJ s}^{-1} \times 60 \text{ s}$$

6 Calculate the answer:

$$Q_{\text{out}} = 8694 \text{ kJ}$$

7 Give the answer to the correct number of significant figures:

$$Q_{\text{out}} = 8700 \text{ kJ}$$

9780170412551

QUESTIONS

1 Determine the efficiency of a system that provides 32 kJ of output energy when 66 kJ energy is put into it.

_____

_____

_____

_____

_____

2 Determine the energy output provided by a 30% efficient system when 23 000 J of energy is put into it.

_____

_____

_____

_____

3 A heat engine operating at its stable operating temperature has an efficiency of 23% and uses 120 kJ of heat energy every second.

a Determine the amount of work that is output by the system every second.

_____

_____

_____

_____

_____

b Determine the amount of heat that is lost to its surroundings in 1.0 minute.

_____

_____

_____

_____

_____

_____

EVALUATION

1 A lawn mower engine is an example of:

A an external combustion engine.

B a heat pump.

C an internal combustion engine.

D a heat-conversion system.

2 Which of the following factors does not appear within the work–energy principle?

A Power

B Heat

C Work

D Internal energy

3 If an engine has a power rating of 13 000 W, how many joules of energy does it produce in 1.0 minute?

A 13 000 J

B 1300 J

C 7800 J

D 780 000 J

4 Which of the following would result in the increase in the internal energy of a system?

A Heat is added to the system.

B Heat is removed from the system.

C Work is done by the system.

D Kinetic energy is removed from the system.

5 An increase in which of the following terms would result in a decrease in the efficiency of a system?

A Q_{in}

B Q_{out}

C W

D U

6 If an engine has its combustion chamber inside of the engine, define what type of combustion engine it is.

_____

7 If heat is being added to a system when no work is being done by the system, explain what must be happening to the internal energy of the system.

_____

8 Use the equations of work and power to show that another way of calculating the power of a system whose internal energy is only changing because it is doing work would be $P = F \times v$, where v is the velocity of an object upon which the force is acting.

_____

_____

_____

9 Explain the thermodynamic processes involved in the operation of an external combustion engine.

_____

_____

_____

10 Explain the thermodynamic processes involved in the operation of a refrigerator.

_____

_____

_____

11 Determine the change in the internal energy of a system when 22 000 J of heat is added to it while 3600 J of work is being done on the system.

_____

_____

_____

_____

12 Determine the amount of work that is being done by an engine when 650 kJ of heat is added to it and 320 kJ of heat is being lost by it, if it is at its stable operating temperature.

_____

_____

_____

_____

_____

13 Determine the efficiency of a system that produces 35 000 J of usable energy when 38 000 J of energy is added to it.

_____

_____

_____

_____

_____

14 If a steam engine that operates at its stable operating temperature has an efficiency of 32%, determine the amount of heat it loses to its surroundings given that it produces 5.6 kJ of useful work.

_____

_____

_____

_____

_____

_____

_____

6 Nuclear model and stability

LEARNING

Summary

▶ The nucleus of an atom contains protons and neutrons, and hence most of the atom's mass.

▶ Electrons are in orbitals around the nucleus of each atom.

▶ Proton and electron charge is equal in magnitude but opposite in charge.

▶ Like charges repel due to electrostatic force.

▶ Elements are defined by the number of protons, isotopes are defined by the number of neutrons within an element, and nuclides are defined by both the number of protons and neutrons, as well as the energy state of the nucleus of an atom.

▶ Standard international notation for an isotope is $_{Z}^{A}X$, where A is the mass number (number of nucleons in the nucleus), Z is the atomic number (number of protons in the nucleus) and X is the element symbol of the isotope.

▶ Atomic weight is an average weight of an element calculated using the percentage abundance of each isotope of that element.

▶ The strong nuclear force is one of the four fundamental forces; the other three are the electromagnetic, gravitational and weak forces.

▶ Strong nuclear force occurs between subatomic particles due to the exchange of mesons.

▶ Nuclear stability occurs when there are particular ratios between neutrons and protons in a nucleus.

▶ The line of stability represents all stable nuclides.

▶ A nucleus is stable if it does not need to undertake radioactive decay to be in its ground energy state.

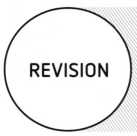

REVISION

6.1 Isotopes

WORKED EXAMPLE

a Pure boron contains 80.22% of isotope $^{11}_{5}\text{B}$ and 19.78% of isotope $^{10}_{5}\text{B}$. Find the atomic weight of boron.

b Antimony has a relative atomic mass of 121.855; 57.25% of this sample is comprised of antimony-121, and the rest is comprised of another isotope of antimony. Determine the mass number of the remaining isotope in this sample.

ANSWER

a Weighted average $= 11 \times 0.8022 + 10 \times 0.1978$ (multiply the mass number by the percentage, and sum the two)

Weighted average $= 10.80$

b The atomic weight of antimony is 121.855, consisting of 57.25% of antimony-121 and 42.75% of an unknown number.

We can express this unknown mass number as x and solve as follows:

$121.855 = 121 \times 0.5725 + x \times 0.4275$

$121.855 = 69.2725 + 0.4275x$

$x = \dfrac{52.5825}{0.4275}$

$x = 123$

The mass number of the unknown isotope of antimony in the sample is 123.

QUESTIONS

1 Distinguish between an isotope and an element.

_____

_____

_____

2 Identify and describe the three isotopes of hydrogen.

_____

_____

_____

9780170412551

3 Write the notation for the three isotopes of hydrogen: hydrogen-1, hydrogen-2 (deuterium) and hydrogen-3 (tritium).

_____

_____

_____

4 A sample of pure potassium contains 93% of $^{39}_{19}K$ and 7% of $^{41}_{19}K$. What is the relative atomic mass of potassium according to this natural abundance of isotopes?

_____

_____

_____

5 The atomic weight of chlorine is 35.45. Within a sample of chlorine, 76% is found to be the isotope chlorine-35, and the remainder consists of a different chlorine isotope. Determine the unknown isotope of chlorine in the sample.

_____

_____

_____

_____

_____

6.2 Nuclides

WORKED EXAMPLE

a State the number of nucleons, protons and neutrons in these isotopes of hydrogen.

  i $^{3}_{1}H$

  ii $^{13}_{6}C$

b State the nuclide with:

  i 14 nucleons and 7 protons

  ii 20 protons and 23 neutrons.

ANSWER

Recall: $^{A}_{Z}X$, where A is the mass number (number of nucleons in the nucleus), Z is the atomic number (number of protons in the nucleus) and X is the element symbol.

a **i** The number of nucleons (A) is 3, and the number of protons (Z) is 1. The number of neutrons (N) is the difference between A and Z.

$$N = A - Z$$
$$N = 3 - 1$$
$$N = 2$$

There are 2 neutrons in this isotope of hydrogen.

  ii The number of nucleons (A) is 13, where the number of protons (Z) is 6. The number of neutrons (N) is the difference between A and Z.

$$N = A - Z$$
$$N = 13 - 6$$
$$N = 7$$

There are 7 neutrons in this isotope of hydrogen.

b For both **i** and **ii**, the final notation will be in the form $^{A}_{Z}X$, so the number of nucleons and protons needs to be determined. Once Z is determined, the periodic table can be used to find the element symbol.

  i 7 protons means the element is nitrogen. We can write $^{14}_{7}N$.

  ii 20 protons means the element is calcium. The addition of the number of neutrons and protons will give mass number A (number of nucleons).

$$A = N + Z$$
$$A = 23 + 20$$
$$A = 43$$

We can now write this nuclide as $^{43}_{20}Ca$.

9780170412551

QUESTIONS

1 State the number of nucleons, protons and neutrons in each of the following nuclides.

a $^{14}_{6}C$

_____

b $^{41}_{19}K$

_____

c $^{7}_{3}Li$

_____

d $^{63}_{29}Cu$

_____

e $^{24}_{12}Mg$

_____

f $^{235}_{92}U$

_____

g $^{78}_{36}Kr$

_____

2 State the nuclides that have the following properties.

a 85 protons, 125 neutrons

_____

b 188 nucleons, 76 protons

_____

c 25 neutrons, 45 nucleons

_____

d 70 nucleons, 38 neutrons

_____

e 84 neutrons, 60 protons

WORKED EXAMPLE

If the distance between two protons is tripled, what would happen to the electrostatic force between the protons?

ANSWER

1 State Coulomb's Law for the force prior to increasing the distance between the protons.

$$F_1 = \frac{kqQ}{r^2}$$

2 The distance has tripled, so the new force will be calculated by:

$$F_2 = \frac{kqQ}{(3r)^2}$$

$$F_2 = \frac{kqQ}{9r^2}$$

$$F_2 = \frac{1}{9} \times \frac{kqQ}{r^2}$$

$$F_2 = \frac{1}{9} \times F$$

Therefore, if the distance between the protons triples, the force will decrease by a factor of nine.

QUESTIONS

3 The distance between two protons in a nucleus is d, and they exert an electrostatic force (F) of repulsion on each other. By what factor would the repulsive force change if the distance between the protons was halved?

_____

_____

_____

_____

_____

9780170412551

6.3 The stability curve

The stability curve (Figure 6.3.1) shows all the stable isotopes of an element as data points, where each data point represents the number of both protons and neutrons within the nucleus. If a nuclide exists with p protons and n neutrons and this data point is *not* on the stability curve, then it can be concluded that the nuclide is *unstable*, and likely to undergo radioactive decay.

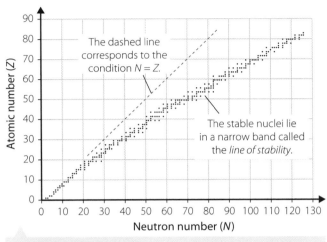

FIGURE 6.3.1 The stability curve

QUESTIONS

1 A neutron to proton ratio can be found by dividing the number of neutrons (N) by the number of protons (Z). According to Figure 6.3.1, there are three table isotopes of neon. List these isotopes and their N:Z ratios.

_____

_____

_____

2 As the atomic number increases, the N:Z ratio increases as well. Suggest an explanation for this.

_____

_____

_____

_____

3 State whether the following isotopes will be stable according to the stability curve.

a $^{127}_{53}\text{I}$

_____

b Nickel-63

_____

c The isotope with 73 protons and 177 nucleons

_____

d Rubidium isotope with 48 neutrons

_____

1 The nucleus of an atom stays together because:

A protons do not repel each other.

B neutrons increase the space between protons.

C of the strong nuclear force.

D masses attract each other.

2 The element represented by $^{112}_{55}$Cs is:

A copernicium.

B magnesium.

C lanthanum.

D caesium.

3 Calculate the relative atomic mass of pure sulfur, given that a sample contains 95% of sulfur-32, 4.45% of sulfur-34 and 0.75% of sulfur-33.

_____

_____

_____

_____

_____

4 Two protons are placed in close proximity to each other. The force of electrostatic repulsion is measured to be F. The distance between the two protons has now decreased to a third of the original separation distance. By what factor does this affect the force the protons exert on each other?

_____

_____

_____

_____

_____

5 Explain the stability of a nuclide in terms of the operation of the strong nuclear force over very short distances, electrostatic repulsion and the relative number of protons and neutrons in the nucleus.

_____

_____

_____

_____

_____

LEARNING

Summary

▶ Radioactivity occurs naturally when unstable nuclei rearrange and release both particles and energy to become more stable.

▶ Radioactive decay is spontaneous.

▶ Parent nuclides are unstable and the daughter nuclides resulting from the radioactive decay are more stable.

▶ Alpha radiation occurs when a very large nucleus emits an alpha particle (comprised of two protons and two neutrons) in order to become more stable.
  - Alpha particles have a very low penetrating ability due to their size, but a high ionising ability due to their charge.
  - Alpha decay can be represented as $^A_Z X \rightarrow {}^{A-4}_{Z-2}Y + {}^4_2\alpha$ where A is the mass number, Z is the proton number, X is the element before decay and Y is the element after α decay.

▶ Beta-positive and beta-minus radiation occurs when a nucleus has too many protons or neutrons in the nucleus respectively. The beta particles are ejected from the nucleus so that it becomes more stable.
  - The ejection of a beta-positive particle (positron) decay is modelled as a proton turning into a neutron by ejecting a positron, making the nucleus more stable.
  - The ejection of a beta-minus particle (electron) decay is modelled as a neutron turning into a proton by ejecting an electron, making the nucleus more stable.
  - Beta particles have a moderate penetrating ability due to their small size, and a small ionising ability as they are charged.
  - Beta-positive decay can be represented as $^A_Z X \rightarrow {}^{A}_{Z-1}Y + {}^0_1\beta + \upsilon$, where A is the mass number, Z is the proton number, X is the element before decay and Y is the element after β^+ decay.
  - Beta-minus decay can be represented as $^A_Z X \rightarrow {}^{A}_{Z+1}Y + {}^0_{-1}\beta + \bar{\upsilon}$, where A is the mass number, Z is the proton number, X is the element before decay and Y is the element after β^- decay.

▶ Gamma radiation is very high energy electromagnetic radiation that often accompanies alpha or beta decay.
  - Gamma radiation has a very high penetrating ability, and a very low ionising ability, as this type of radiation has no charge.
  - Gamma decay can be represented as $^A_Z X^* \rightarrow {}^A_Z X + \gamma$, where the mass and atomic number do not change in gamma decay, but rather the nucleus goes from an excited state to its more stable ground state by giving off a large amount of energy.

- The half-life of an isotope is the time it takes for half of a radioactive sample to decay into the daughter nuclide.

- Half-life is modelled as $N = N_0 \left(\dfrac{1}{2} \right)^n$, where for a sample of N_0 particles, the number N remaining after n half-lives can be found.

- Radioactive nuclei will undergo a series of spontaneous decays, to become stable, when the nucleus is finally small enough to have an $N{:}Z$ ratio that is stable (as per the line of stability).

7.1 Types of radiation

QUESTIONS

1 State the charge on an alpha particle, a positron and an electron.

_____

_____

_____

2 Compare the three types of radiation in terms of their penetrating and ionising ability.

_____

_____

_____

3 Explain why some nuclides undergo radioactive decay and others do not.

_____

_____

_____

_____

_____

4 How can you predict what type of decay an unstable nuclide will undergo?

_____

_____

_____

_____

_____

9780170412551

WORKED EXAMPLE

a What is the daughter nuclide produced after the following?

   i Holmium-151 decays by α emission

   ii Fluorine-21 decays by β^- emission

b Balance these decay equations.

   i $^{238}_{92}U \rightarrow \underline{\hspace{1em}} + ^4_2\alpha$

   ii $^{195}_{81}Tl \rightarrow ^{195}_{80}Hg + \underline{\hspace{1em}}$

ANSWER

a **i** Holmium has the atomic number 67. It can be written as $^{151}_{67}Ho$. The daughter nuclide must have $151 - 4 = 147$ nucleons, and $67 - 2 = 65$ protons. This is so that the left- and right-hand sides of the decay equation are conserved under the conservation of energy and mass.

    So now we have $^{151}_{67}Ho \rightarrow ^{147}_{65}\underline{\hspace{1em}} + ^4_2\alpha$

    From the periodic table, the element with atomic number 65 is terbium. Therefore, the daughter nuclide is terbium-147.

  ii Using the same logic as in part (i), we can deduce that fluorine-21 can be written as $^{21}_9F$ and that the daughter nuclide will have the same number of nucleons, but one more proton.

    We now have $^{21}_9F \rightarrow ^{0}_{10}\underline{\hspace{1em}} + ^{0}_{-1}\beta$.

    From the periodic table, the element with atomic number 10 is neon. Therefore, the daughter nuclide is neon-21.

b For both equations, the number of nucleons on the left-hand side needs to equal to the number of nucleons on the right-hand side, and the number of protons on the left-hand side must equal the number of proton on the right-hand side.

   i For $^{238}_{92}U \rightarrow \underline{\hspace{1em}} + ^4_2\alpha$, we know the daughter nuclide must have $238 - 4 = 234$ nucleons, and $92 - 2 = 90$ protons. From the periodic table, the element with atomic number 90 is thorium. Hence, the equation can be balanced as follows: $^{238}_{92}U \rightarrow ^{234}_{90}Th + ^4_2\alpha$.

  ii For $^{195}_{81}Tl \rightarrow ^{195}_{80}Hg + \underline{\hspace{1em}}$, we know that the decay that thallium undertakes does not affect the nucleon number, meaning that the type of decay is beta. In order for the proton numbers to balance, the beta radiation must be a positron. Hence, the equation can be balanced as follows: $^{195}_{81}Tl \rightarrow ^{195}_{80}Hg + ^0_1\beta$.

QUESTIONS

5 Balance the following decay equations.

   a $^{241}_{95}Am \rightarrow \underline{\hspace{1em}} + ^4_2\alpha$

_____

_____

_____

b $^{90}_{38}Sr \rightarrow {}^{90}_{39}Y + \underline{\quad}$

_____

_____

_____

c $\underline{\quad} \rightarrow {}^{11}_{5}C + {}^{0}_{1}\beta$

_____

_____

_____

d $^{\underline{\quad}}_{15}P \rightarrow {}^{30}_{14}\underline{\quad} + {}^{0}_{\underline{\quad}}\beta$

_____

_____

_____

6 State the daughter nuclide for the following decays.

   a Polonium-84 decays via α emission.

_____

_____

   b Lead-210 decays via β^- emission.

_____

_____

   c Radon-220 decays via α emission.

_____

_____

   d Oxygen-15 decays via β^+ emission.

_____

_____

7 State the decay that the parent nuclide has undergone to turn into the respective daughter nuclide.

   a Carbon-14 decays to nitrogen-14.

_____

_____

b Aluminium-26 decays to magnesium-26.

_____

_____

c Bismuth-214 decays to polonium-214.

_____

_____

d Californium-249 decays to curium-245.

_____

_____

7.2 | Half life

WORKED EXAMPLE

Europium-150 has a half-life of approximately 12.8 hours.

a How long will it take for an 80 g sample of europium-150 to only have 20 g of the original isotope remaining?

b What is the decay constant of europium-150?

ANSWER

a 1 First, determine how many half-lives this would take by rearranging the equation to make n the subject:

$$N = N_0 \left(\frac{1}{2} \right)^n$$

$$\frac{N}{N_0} = \left(\frac{1}{2} \right)^n$$

$$n = \log_{\frac{1}{2}} \frac{N}{N_0}$$

2 Substitute in values:

$$n = \log_{\frac{1}{2}} \frac{20}{80}$$

$$n = 2 \text{ half-lives}$$

Each half-life is 12.8 hours long, and $12.8 \times 2 = 25.6$ hours.

Therefore, it will take 25.6 hours for only 20 g of europium-150 to remain.

b **1** The decay constant is directly related to the half-life of an isotope:

$$t_{\frac{1}{2}} = \frac{\ln 2}{\lambda}$$

2 Rearrange for λ

$$\lambda = \frac{\ln 2}{t_{\frac{1}{2}}}$$

$$\lambda = \frac{\ln 2}{12.8}$$

$$\lambda = 0.0541 \text{ decays per hour}$$

QUESTIONS

1 How long will it take a sample of bismuth-214 to only have 12.5% of the original nuclei remaining if the half-life of this isotope is 20 minutes?

_____

_____

_____

2 Determine how many half-lives have elapsed for a sample of radioactive neodymium if $\frac{1}{16}$ of the original nuclei remaining.

_____

_____

_____

3 How much of a 150 g sample of radioactive radium remains after 7 half-lives have elapsed?

_____

_____

_____

4 The activity of a sample is directly proportional to the amount of the original radioactive sample remaining. If the half-life of a radioactive isotope of carbon is approximately 5700 years, and the original activity of this sample was 100 counts per minute, how long will it take for the activity to fall to 10 counts per minute?

_____

_____

_____

9780170412551

1 A pure sample of 1024 radioactive nuclides decays for 3 half-lives. The number of nuclides remaining is:

A 103

B 128

C 0.125 108

D 0.125 102 4

2 $^{211}_{87}$Fr decays via alpha particle emission. The daughter nucleus is:

A $^{211}_{88}$Fr

B $^{207}_{85}$At

C $^{211}_{87}$Fr

D $^{209}_{85}$At

3 In the decay equation $^{57}_{26}$Fe $\rightarrow$ X + γ, what is X?

A $^{58}_{26}$Fe

B $^{57}_{26}$Fe

C $^{58}_{27}$Co

D $^{57}_{27}$Co

4 Tantalum-181 is stable. Two of its isotopes, tantalium-178 and tantalium-183, decay to stable nuclei after beta emission. Show the decay sequences for each.

a Beta-minus emission

_____

_____

_____

_____

b Beta-plus emission

_____

_____

_____

_____

5 Zinc-72 decays by way of beta emissions to a stable nuclide of germanium. Show the decay sequence.

6 A sample of an unknown radioactive material was analysed every hour over a period of time by measuring the level of radiation emitted by the sample. The data was collected and is displayed in Table 7.3.1.

TABLE 7.3.1 Radiation measured after t hours from a radioactive source

TIME (hours)	RADIATION (counts per second)
0	7932
1	5430
2	3720
3	2545
4	1740
5	1190
6	820
7	560
8	385
9	260
10	180
11	124
12	83
13	58
14	40

9780170412551

Plot the results from Table 7.3.1 as radiation intensity against time on a set of axis that is labelled appropriately. Draw a curved line of best fit for this data. Using at least three points of data from your graph, approximate the half-life of this unknown material.

_____

_____

_____

_____

7 The term 'radionuclide' is more precise than 'radioisotope'. Explain why. (This is why 'radionuclide' has now mostly replaced the term 'radioisotope' within nuclear science.)

_____

_____

_____

_____

LEARNING

Summary

▸ Energy can be produced from nuclei through nuclear fission and nuclear fusion.

▸ Nuclear fission is the process by which nuclei separate into fragments, with the release of energy.

▸ Nuclear fusion is the process by which light nuclei are fused together, with the accompanying release of energy.

▸ There are four fundamental forces that act within a nucleus, with different comparative strengths. They include the strong nuclear force, the weak nuclear force, the electromagnetic force and the gravitational force.

▸ The binding energy per nucleon governs stability; it is the energy required to disassemble a nucleus into its component nucleons.

▸ The mass defect, Δm, is the difference between the mass of an atom and the mass of its constituent parts, as expressed in Einstein's mass–energy equation: $E = \Delta mc^2$.

▸ Nuclear fission is the process by which heavy nuclei separate into fragments, with the accompanying release of energy.

▸ A fission chain reaction occurs when more than one of the neutrons released from the initial fission event causes new events to occur.

▸ Nuclear energy production by fission reactors produces radioactive waste, the disposal of which is a significant issue.

▸ Fusion reactions release much more energy than fission reactions per kilogram of reactant.

9780170412551

8.1 Nuclear energy and mass defect key terms

QUESTIONS

1 Demonstrate your understanding of the energy released in a nuclear fission reaction by writing a paragraph to relate the key terms below.

- **mass defect, Δm** – the difference between the mass of an atom and the mass of its constituent parts, as expressed in Einstein's mass–energy equation: $E = \Delta mc^2$

- **Einstein's mass–energy equivalence equation** – $E = \Delta mc^2$, where $c =$ the speed of light, $3.00 \times 10^8\,\mathrm{m\,s^{-1}}$

- **fission chain reaction** – occurs when the neutrons released from a fission event go on to produce more fission events. Controlled fission reactions are used in thermal nuclear reactors to generate energy. Uncontrolled fission reactions are used in weapons

- **controlled nuclear chain reaction** – a chain of nuclear reactions that are controlled to limit the rate at which reactions occur. In a steady state (reaction rate held constant), an average of one neutron from each reaction goes on to cause another reaction. This is the case for a nuclear power reactor running at constant power output

- **fission fragment** – nucleus produced as a result of fission; fission product

- **slow (thermal) neutron** – neutron with kinetic energy around 0.1–20 keV

_____

_____

_____

_____

_____

_____

_____

_____

_____

_____

_____

_____

_____

_____

2 Draw a flow chart or mind map to relate and link the following key terms.

- **nucleon** – a component of the nucleus; a proton or a neutron

- **strong nuclear force** – the force required to hold nucleons together, especially to overcome the electrostatic force of repulsion between protons

- **fission** – the process by which heavy nuclei ($Z > 56$) separate into fragments, with the release of energy. Typically, fission fragments have quite different masses

- **fusion** – the process by which nucleons join to form a new nucleus. Nucleosynthesis is the set of fusion reactions that lead from nucleons to a variety of nuclides. This occurs for light elements ($Z < 56$) and energy is released

- **electromagnetic force** – the combined electrical and magnetic force acting between charged particles. The force is attractive for unlike charges, and repulsive for like charges

- **gravitational force** – the manifestation of Newton's universal law of gravitation; a force of attraction acting between every mass throughout the universe

- **weak nuclear force** – one of the four fundamental forces; it acts between subatomic particles (leptons) and is responsible for beta decay

9780170412551

8.2 | Energy source evaluation

QUESTIONS

You have been appointed head of the Australian Federal Government's 'Energy for the Future' advisory committee. This committee is about to recommend the types of energy resources that should be used in the coming decades to provide a reliable, clean, cheap and safe source of electricity for the country.

Using nuclear power and any two of coal-fired, hydro, geothermal, wind or solar power, perform research to complete the information in a table that compares the appropriateness of the three chosen energy sources across a range of criteria for the purpose of the report to the government.

The criteria for evaluation in the table can include the following factors: availability of the resource in Australia, running cost (per MWh), initial cost, reliability and suitability in various climates and weather conditions, greenhouse gas emissions and safety.

Finally, in a written summary, propose which of your three sources of energy you would advise the government to proceed with as Australia's preferred energy source for the next 50 years.

TABLE 8.2.1 Energy source evaluation

CRITERIA	NUCLEAR POWER	FUEL SOURCE 2	FUEL SOURCE 3
Source availability			
Running cost (per MWh)			
Initial cost			
Reliability and suitability			
Greenhouse gas emissions			
Safety			

_____

_____

_____

_____

_____

_____

8.3 | Energy within the nucleus

QUESTIONS

1 The strong nuclear force overcomes the electrostatic force of repulsion between protons at close range, typically femtometres ($\times 10^{-15}$ m) within the nucleus. The electrostatic force can be found using Coulomb law, $F = \dfrac{kqQ}{d^2}$, where $k = 9.0 \times 10^9 \, \mathrm{N\,m^2\,C^{-2}}$, q and Q are the magnitude of the charges ($+1.60 \times 10^{-19}$ C) and d is the distance (in metres) between the charges.

   a Find the magnitude of the force of electrostatic repulsion between two protons inside a nucleus when their centres are separated by the following distances.

      i 2 femtometre (fm)

      ii 4 femtometre (fm)

   b State the magnitude of the strong nuclear force at 2 fm.

2 Neutrons are able to assist in moderating the electrostatic force of protons on other protons. State how it is that neutrons may assist in reducing the proton–proton repulsion.

3 The first eight elements have stable nuclides for $N = Z$, apart from beryllium-9, which is the only stable nuclide of beryllium. The others have two stable nuclides, except oxygen, which has three. Except for $^1_1\mathrm{H}$ and $^3_4\mathrm{He}$, in all the stable nuclides the number of neutrons is equal to or greater than the number of protons.

   a Place all these stable nuclides on the stability chart of atomic number, Z, versus number of neutrons, N (Table 8.3.1). Three examples have been provided for you.

   b Tritium is an unstable nuclide of hydrogen. Place this on the chart.

9780170412551

c Carbon nuclides range from $A = 9$ to $A = 16$. Place these on the chart.

TABLE 8.3.1 Stability chart of light elements

Number of neutrons, N	1	2	3	4	5	6	7	8
16								0–16
15								
14								
13								
12								
11								
10								
9								
8								
7								
6								
5								
4								
3		He-3						
2								
1								
0	H-1							

Atomic number, Z

4 Show that $1\,u = 1.66 \times 10^{-27}\,kg$ is equivalent to $931.3\,MeV$.
(Hint: Use $E = mc^2$ and the conversion between joules and electron-volts, $1\,MeV = 1.60 \times 10^{-13}\,J$.)

_____

_____

_____

_____

_____

_____

5 Complete the table of masses of subatomic particles.

TABLE 8.3.2 Masses of subatomic particles

PARTICLE	MASS		
	kg	u	
Proton	1.673×10^{-27}		
Neutron		1.008 67	
Electron			0.511

8.4 | The four fundamental forces

QUESTIONS

1   Complete the headings of Table 8.4.1 for the four fundamental forces acting within a nucleus. Select from gravitational force, electromagnetic force, strong nuclear force, and weak nuclear force.

TABLE 8.4.1 The four fundamental forces

	TYPE OF FORCE			
Relative magnitude	1	10^{32}	10^{36}	10^{40}
Range (m)	Infinite	10^{-18} or 1 attometre, 1 am	Infinite	10^{-15} or 1 femtometre, 1 fm

2   Complete the following sentences.

   a   _____ have a positive charge.

   b   _____ charges repel, whereas _____ charges attract.

   c   The repulsion that exists between protons within the nucleus is the result of _____ force.

   d   The force that keeps nucleons together is termed the _____ _____ force.

   e   Einstein's mass–energy equation is quantitatively stated as: _____.

3   Using the data provided for an alpha particle (the $^4_2$He nucleus), determine the following:

   a   The mass defect, in kg

   b   The binding energy per nucleon (MeV)

   (Hint: an alpha particle has 2 protons, 2 neutrons and no electrons).

   Unified atomic mass unit, u = 1.66×10^{-27} kg

   Rest mass of proton, m_p = 1.007 28 u or $1.672\ 08 \times 10^{-27}$ kg

   Rest mass of neutron, m_n = 1.008 66 u or $1.674\ 38 \times 10^{-27}$ kg

   Rest mass of alpha particle, m_α = 4.001 53 u or $6.642\ 54 \times 10^{-27}$ kg

   _____

   _____

   _____

   _____

4   Complete the missing values in the nuclear fusion equations.

   a $^2_1H + {}^2_1H \rightarrow {}_2He + {}^1_0n + energy$

   b $^2_1H + {}^3H \rightarrow {}^4_2He + {}^1_0n + energy$

5   Distinguish between a 'slow' or thermal neutron and a 'fast' neutron.

   _____

   _____

   _____

9780170412551

6 Draw a diagram to illustrate a nuclear chain reaction.

8.5 | Radioactive waste management

QUESTIONS

1 ANSTO, the Australian Nuclear Science and Technology Organisation, describes nuclear waste as falling into three categories. Identify these categories and give an example of each.

_____

_____

_____

_____

_____

_____

2 Nuclear waste must be managed in accordance with strict standards, which are set both internationally and domestically.

a Which organisation is the international regulator?

_____

b Which organisation is the Australian regulator?

_____

3 What units are used for measuring radiation?

_____

4 How much nuclear waste, by category, is currently stored in Australia?

_____

_____

_____

5 How many sites in Australia are storing nuclear waste?

_____

6 Consider the statement that 'Australia should have a main storage site for most of its nuclear waste'. Provide an argument to support this statement.

_____

_____

_____

_____

7 What phenomenon causes radioactive waste to become less radioactive over time, at different rates for different substances? Include a diagram in your response, to explain the process.

_____

_____

_____

_____

8 Recall the characteristics and penetrating power of the different types of radiation to complete Table 8.5.1.

TABLE 8.5.1 Nuclear radiation characteristics

RADIATION TYPE	PHYSICAL ENTITY	CHARGE	RELATIVE PENETRATION ABILITY	ENVIRONMENTAL IONISATION
Alpha				
Beta				
Gamma				

9 High-level waste requires long-term storage. What geological and environmental factors would be needed for a high-level waste repository?

_____

_____

_____

9780170412551

10 Evaluate this statement: 'Australia is in an ideal location to have a long-term high-level waste storage facility, providing economic benefits for Australia and a storage solution for the world.'

_____

_____

_____

_____

_____

8.6 | Safety in nuclear reactors

QUESTIONS

1 Research to find what, according to the United States Nuclear Regulatory Commission, are three important concerns of safety-related functions.

_____

_____

_____

2 Perform research to complete Table 8.6.1 regarding nuclear accidents at nuclear plants globally.

TABLE 8.6.1 Nuclear plant accidents

YEAR	NUCLEAR PLANT	DETAILS OF THE EVENT	APPROXIMATE NUMBER OF FATALITIES
1979	Three Mile Island, United States		
1986	Chernobyl, Ukraine		
2011	Fukushima, Japan		

3 Research the Pike River (New Zealand) mine disaster, as well as the death rate from coal mining in the United States, United Kingdom and China. Evaluate the safety of nuclear power versus coal power, including that of both employees and the population at large.

_____

_____

_____

_____

_____

_____

_____

_____

4   Perform research into reactor protection systems to answer the following questions.

   a   How does a reactor protection system function?

   _____

   _____

   b   What are control rods?

   _____

   _____

   c   What is the purpose of the safety injection or standby liquid control system (SLCS), and how is it
        accomplished?

   _____

   _____

   d   Why is boron an important element in this system?

   _____

5   **a**   What is the emergency core cooling system (ECCS)?

   _____

   _____

   _____

   b   Under what conditions is an ECCS activated?

   _____

6   Describe the structure and function of the following ECCS components.

   a   High-pressure coolant injection system

   _____

   _____

   b   Automatic depressurisation system

   _____

   _____

   c   Low-pressure coolant injection system

   _____

   _____

   d   Core spray system

   _____

   _____

e Containment spray system

_____

_____

f Isolation cooling system

_____

_____

7 Describe the structure and functions of the following elements of a containment system.

a Fuel cladding

_____

_____

b Reactor vessel

_____

_____

c Primary containment

_____

_____

d Core catching

_____

_____

8 What do most nuclear plants do to reduce the risk of radiation exposure from the air if a radiation release has occurred?

_____

_____

_____

_____

EVALUATION

1   Of the four fundamental forces, which one is associated with overcoming the electrostatic repulsion of protons within the nucleus?

   A   Weak nuclear force

   B   Strong nuclear force

   C   Gravitational force

   D   Electromagnetic force

2   Which form of nuclear radiation or emission has medium penetration ability, being absorbed by aluminium a few millimetres thick, and is a form of ionising radiation that behaves like a negative charge?

   A   Beta-minus emission

   B   Beta-positive emission

   C   Alpha emission

   D   Gamma radiation

3   Einstein's mass–energy equivalence equation relates the properties E, Δm and c as $E = mc^2$. The correct units for the variables E, m and c respectively are:

   A   megajoules, grams, kilometres per hour.

   B   joules, kilograms, metres per second squared.

   C   kilojoules, grams, metres per second.

   D   joules, kilograms, metres per second.

4   Define the term 'ionising radiation'.

   _____

   _____

5   How can beta-minus emission result in a more stable nucleus?

   _____

   _____

6   Outline the potential advantages of nuclear fusion power generation over nuclear fission power generation.

   _____

   _____

   _____

   _____

7 What are the differences between the moderator and the control rods in a thermal nuclear reactor?

_____

_____

_____

8 Explain how protons are kept together in the nuclei of atoms despite the force of electrostatic repulsion between them.

_____

_____

9 Explain how you would make a model of an atom, such as helium, that shows the position of the particles and their relative sizes.

_____

_____

10 Explain how the binding energy for a nucleus can be calculated.

_____

_____

11 Using a diagram, compare the processes of fusion and fission.

Fusion

Fission

12 Outline the events that would lead to an uncontrolled fission chain reaction of uranium.

_____

_____

_____

13 Why is harnessing the power from controlled nuclear fusion to generate electricity still years or even decades away?

_____

_____

14 Australia has not embraced nuclear power; however, many other countries continue to depend on it for their energy needs. If the Australian Federal Government decided to build two new nuclear power stations, where geographically would you propose that they be built? Explain your reasoning.

_____

_____

_____

_____

_____

_____

9780170412551

Current, potential difference and energy flow

LEARNING

Summary

- Electricity is a convenient and versatile form of energy that is available from a variety of sources.

- Electrical energy may be supplied by either direct current (DC) or alternating current (AC) sources.

- The Rutherford–Bohr model of the atom contains a central, positively charged nucleus surrounded by negatively charged electrons in discrete electron energy levels.

- Electric charge is conserved; it cannot be created nor destroyed.

- Like charges repel; opposite charges attract.

- Electric charge can be transferred from one object to another. Separated charge is stored as electrical potential energy, for example in a battery.

- An electric circuit is a conducting path within which electrons may flow. The flow of charge is termed an electric current and is measured using an ammeter, connected in series.

- Insulators do not allow charges to flow, whereas conductors allow electrons to move freely within the material.

- Kirchhoff's current law asserts that the total current arriving at a junction within an electrical circuit is equal to the total current leaving the junction.

- The potential difference between any two points in an electric circuit can be determined using a voltmeter, connected in parallel.

- Kirchhoff's voltage law asserts that for any closed path in an electric circuit, the sum of the potential differences adds to zero.

- The SI unit for energy is the joule; however, the kilowatt hour (kWh) is more frequently used for measuring electricity consumption.

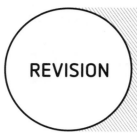
9.1 Current, potential difference and energy flow key terms

1 Demonstrate your understanding of electric circuits by writing a paragraph to relate the key terms below.

- **conventional current** – the convention to describe electrical current as the flow of positive charge

- **conductor** – a material that allows the flow of electrons; e.g. metals

- **current, I** – the rate of flow of charge, that is, charge per unit time; measured in amperes, A: $I = \dfrac{q}{t}$

- **direct current (DC)** – current that flows in one direction in a conducting path

- **electric circuit** – a complete conducting loop through which elementary charges can flow

- **electron** – a negatively charged subatomic particle and the primary charge carrier in conductors; $q_e = -1.60 \times 10^{-19}\,C$

- **metal lattice** – a regular arrangement of large numbers of metal atoms that allows free electrons to move readily

- **potential difference, V** – potential energy per charge, voltage: $V = \dfrac{W}{q}$

_____

_____

_____

_____

_____

_____

_____

_____

_____

_____

_____

2 Draw a flow chart or mind map to relate and link the following key terms.

- **electron** – a negatively charged subatomic particle and the primary charge carrier in conductors; $q_e = -1.60 \times 10^{-19}\,\text{C}$

- **proton** – a positively charged subatomic particle found within the atomic nucleus; $q_p = +1.60 \times 10^{-19}\,\text{C}$

- **neutron** – a neutral subatomic particle found within the atomic nucleus. Its mass is approximately the same as that of a proton

- **law of conservation of charge** – the net charge within an isolated system is constant

- **static electricity** – charges at rest, or stationary on an object; typically produced on insulators by friction between two surfaces

9.2 | Charge carriers and static electricity

1 Explain how rubbing two different materials together can cause a build-up of electrostatic charge.

_____

_____

2 Suggest a procedure that could be followed to show that like charges repel.

_____

_____

_____

_____

3 Draw a sketch to show the arrangement of electrons in a conductor that allows them to conduct electricity freely.

4 Contrast electrons, protons and neutrons with regard to charge and mass.

_____

_____

_____

5 Explain the difference between static and current electricity.

_____

_____

6 Provide three common examples of static electricity phenomena.

_____

_____

_____

_____

9780170412551

9.3 | Current electricity and electrical devices

1 Complete the passage by filling in the missing words from the list provided.

heat	watts	generators	circuits	gap	chemical reactions
light	globes	current	power	energy	

Electric _____ provide a pathway for _____ to flow. Any _____ in the circuit will prevent this from happening. Devices such as _____ convert electrical _____ into other forms of energy such as _____ or _____.

Electrical energy can be produced using _____ or _____. The rate at which electrical energy is being used is termed _____, which is measured in _____, named after a famous British scientist.

2 Complete Table 9.3.1 to summarise the characteristics of electrical devices.

TABLE 9.3.1 Characteristics of electrical devices

DEVICE	SYMBOL	FUNCTION OF DEVICE
Battery		
Voltmeter		
Switch		
Globe		
Resistor		
Ammeter		

3 Electricity is taken for granted throughout our society; however, its misuse can cause death. Outline the safety features that are found in a house and in appliances that make electricity safe to use.

_____

_____

_____

_____

_____

4 Contrast 'energy' and 'power' by writing a brief paragraph that summarises the terms.

_____

_____

_____

_____

5 a Use a household electricity bill to determine the unit of energy and the typical cost of electricity sold to consumers.

_____

b Convert this unit into joules, given that 1 watt = 1 joule per second, and 1 kW = 1000 W.

_____

_____

6 Why does electricity need a 'circuit' to flow around?

_____

_____

7 How does a switch turn a globe off?

_____

_____

8 Complete the following sentences.

Electrical energy can be converted into other forms of energy such as _____ or _____.

The battery provides a source of _____ for an electric circuit.

The rate at which electricity flows around a circuit is called the electric _____.

9 List five different appliances that use electricity. For each appliance, identify the form (or forms) of energy to which electrical energy is converted.

_____

_____

_____

_____

_____

10 Outline how the use of electricity, electric circuits and batteries have benefitted modern society.

_____

_____

_____

_____

9780170412551

9.4 | Electrical circuit symbols

Draw the circuit symbols for the following electrical devices.

TABLE 9.4.1 Electrical circuit symbols

ELECTRICAL DEVICE	CIRCUIT SYMBOL
Switch (open)	
Fixed resistor	
Lamp	
LED	
Battery of cells	
Ammeter	
Voltmeter	
AC supply	
Cell	
Earth	

9.5 | Series and parallel circuits

1 a What is the name of the law that relates the three quantities voltage, resistance and current?

_____

b Write down the relationship between the resistance of a globe, R, and the current flowing through it, I, when a constant voltage, V, is applied.

_____

_____

c Explain why this is regarded as an 'inverse' relationship?

_____

_____

2 Contrast a series circuit with a parallel circuit. Use an illustration to assist your explanation.

_____

_____

3 A circuit for vehicle lights uses two pairs of 12-volt light globes: a pair of 12 watt globes that provide parking lights and a pair of 24 watt globes that provide brake lights, which are brighter than the parking lights. Draw the vehicle lighting circuit and determine the current running through each parallel circuit.

_____

_____

_____

_____

4 Incandescent light globes use a small filament of wire that glows due to its high resistance. Such globes are approximately 5% efficient in transforming electrical energy into light. Modern LED globes are approximately 90% efficient. Calculate how many times brighter LED globes would be compared with incandescent globes, if placed in the same circuit.

_____

_____

_____

5 Car headlights can be wired to the battery using either series or parallel circuit designs. State which circuit is preferable, and justify this statement.

_____

_____

_____

9.6 DC and AC circuits

1 Contrast direct current (DC) with alternating current (AC).

_____

_____

_____

2 Draw the circuit symbols used for both DC and AC.

3 List three devices that use DC electricity.

_____

4 State the potential difference and frequency of the standard AC mains power supply in Australia.

_____

9.7 Electrical circuit analysis

1 A charge of 30 C flows through the resistor R shown in Figure 9.7.1 over a period of 1 minute.

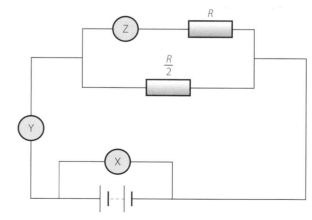

FIGURE 9.7.1 Circuit analysis

a Calculate the current in the resistor R and state the unit.

_____

_____

b Name the three meters and state the quantities that they measure.

X: _____

Y: _____

Z: _____

c If the potential difference across the resistor R is 3.0 V, calculate the resistance R.

_____

_____

d State the reading on each meter.

X: _____

Y: _____

Z: _____

2 An electric kettle is designed to be operated from a 240 V mains supply. When switched on, a current of 6 A flows in the kettle.

a Calculate the power rating of the kettle.

_____

_____

b Calculate the resistance of the heater element.

_____

_____

c Suggest a rating (maximum current, in amperes) for the fuse that should be used with the kettle.

_____

_____

d When the kettle is operating normally, what current is flowing in:

i the live wire?

_____

ii the earth wire?

_____

9780170412551

e If this kettle is used three times a day, for 7 minutes each time, and the cost of the electrical unit is 20 cents per kWh, calculate the cost of using this kettle from 1 January to 31 March in a leap year.

_____

_____

_____

_____

3 A light globe contains a thin tungsten filament surrounded by a noble gas such as argon.

a Explain why the filament is thin.

_____

_____

b State the properties of tungsten and explain why it is used as the filament.

_____

_____

_____

c Explain why a noble gas is used to surround the filament.

_____

_____

d State the energy changes that take place when such a lamp is used.

_____

_____

9.8 | Electrical appliances

1 If a length of resistance wire is connected to the terminals of a cell, which of the following changes will decrease the current in the wire?

   A Increasing the length of the wire with the same cross-sectional area

   B Applying a higher voltage

   C Connecting an identical wire in parallel with the first one

   D Covering the wire with plastic insulation

2 In an experiment, a 3.0 A current flows through a component for a period of 3 minutes. The voltage across the component is measured as 12 V. What quantity of charge flows through the component during the experiment?

   A 108 C

   B 9 C

   C 36 C

   D 540 C

Refer to the information in Table 9.8.1 to answer to the following questions.

TABLE 9.8.1 Electrical appliance characteristics

	APPLIANCE	POWER (W)	VOLTAGE (V)
A	Washing machine	3000	240
B	Car headlight	36	12
C	Television	100	240
D	Electric iron	960	240
E	Hair curling iron	20	150

3 Which appliance has the largest working current?

_____

4 Which appliance has the largest working resistance?

_____

5 The hair curling iron is used for 7 minutes. The energy used is:

   A 140 J

   B 8.4 kJ

   C 1050 J

   D 21 kJ

9780170412551

6 Two resistors R and $2R$ are connected in series as shown in Figure 9.8.1. The ammeters W and Z are connected in series. The voltmeters X and Y are connected in parallel to the respective resistors in the circuit.

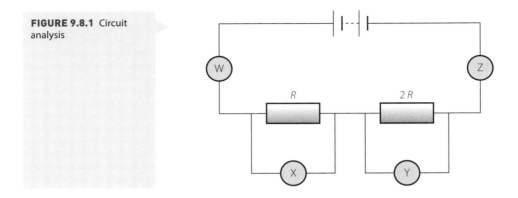

FIGURE 9.8.1 Circuit analysis

Which of the following correctly describes the ammeter and voltmeter readings?

	AMMETER READINGS	VOLTMETER READINGS
A	W is less than Z.	X is less than Y.
B	W is less than Z.	X is greater than Y.
C	W is equal to Z.	X is less than Y.
D	W is equal to Z.	X is equal to Y.

9 In the circuit in Figure 9.8.2, if the ammeter reading is 0.5 A and it has a negligible resistance, what is the value of the resistor R?

FIGURE 9.8.2 Resistors in series

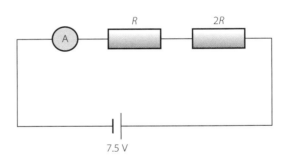

A $1.5 \, \Omega$

B $0.5 \, \Omega$

C $5 \, \Omega$

D $7 \, \Omega$

EVALUATION

1 Answer *true* or *false*. If the statement is false, re-write the statement to make it true.

a Energy is the ability to do work on the surroundings.

_____

b Power is the rate at which energy is expended.

_____

c A 15-watt light globe uses 15 joules of energy every hour.

_____

d When a vehicle doubles its speed, its kinetic energy doubles.

_____

e An example of potential energy is chemical energy stored in a fuel.

_____

f Energy is always lost when converted from one form to another.

_____

g The unit of energy is the joule. The unit of power is the watt, which is one joule per second.

_____

h Energy can take many forms including electrical, chemical, elastic and heat energy.

_____

i Friction sometimes produces heat energy.

_____

2 Given that $W = F \times s$, calculate the distance through which a car is moved by a force of 500 N when 25 kJ of work is done.

A 12 500 m

B 50 m

C 0.05 m

D 20 m

9780170412551

3 How much kinetic energy does a 100-tonne asteroid travelling at $8.0\,\mathrm{km\,s^{-1}}$ possess?

    A $3200\,\mathrm{J}$

    B $3.2 \times 10^{12}\,\mathrm{J}$

    C $3\,200\,000\,\mathrm{J}$

    D $4.0 \times 10^{8}\,\mathrm{J}$

4 Which vehicle has greater kinetic energy: a 25-tonne truck travelling at $5.0\,\mathrm{m\,s^{-1}}$ or a 2000 kg car moving at $30\,\mathrm{m\,s^{-1}}$? Support your response with calculations.

_____

_____

5 Sketch an electric circuit containing a battery and two light globes in series. Include an ammeter and a voltmeter to measure values for one of the globes.

6 An ion drive rocket motor on a spacecraft can exert a small force for a long time. Explain why such a motor could be used for long-distance space travel in the future.

_____

_____

7 Explain the relationships between work, energy, power and time.

_____

_____

8 A light globe is connected in an electric circuit. The reading on the voltmeter placed across the globe is 3.0 V and the reading on the ammeter placed in series with the globe is 0.70 A.

   a Determine the power of the globe.

   _____

   _____

   b Calculate the resistance of the globe.

   _____

   _____

9 How long would it take for 3.5 C of charge to flow through a device that has a current of 0.70 A flowing through it?

   _____

   _____

10 Explain why wires that are designed to carry larger currents typically have greater thickness. Hint: Use the formula for resistivity.

   _____

   _____

   _____

11 Electricity is charged at 25 cents per kWh. A fridge motor connected to a 240 V supply uses 0.80 A of current and runs for an average of 4 hours per day. Determine the cost of running the fridge for one year.

   _____

   _____

9780170412551

LEARNING

Summary

- A potential difference applied across a conducting wire produces an electric current. The opposition to this flow is termed resistance.

- A conductor is a material with free electrons that allows current to flow within it. Metals are good examples of conductors.

- An insulator does not allow current to flow through it easily. Plastics and ceramics are examples of insulators.

- A semiconductor is a material that has a small number of free electrons in its lattice structure and that allows current to flow, but not easily. The resistance of a semiconductor varies with temperature.

- Resistivity, ρ, is a measure of the opposition to flow of electric charge. Resistivity is unique to each material and varies with temperature.

- The resistance of a conducting wire varies with respect to its resistivity, length and cross-sectional area, in accordance with the resistivity formula, $R = \rho \dfrac{\ell}{A}$.

FIGURE 10.0.1 The resistance of a length of wire is dependent upon its resistivity, length and cross-sectional area

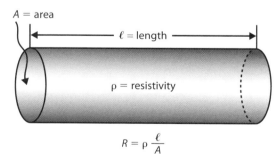

- The resistance, R, of a circuit component is defined as the ratio of the potential difference, V, to the current, I, that is, $R = \dfrac{V}{I}$, often written as $V = IR$.

- An ohmic device has a constant resistance for a wide range of voltages and currents; a non-ohmic device does not have a constant resistance. The current–voltage graph for an ohmic resistor is linear.

10.1 Revising resistance

1   Demonstrate your understanding of ohmic and non-ohmic materials by writing a paragraph to relate the key terms below. Sketch a graph to assist in your explanation.

- **non-ohmic device** – a component that does not provide a constant resistance: $R \neq \dfrac{V}{I}$
- **Ohm's law** – the current that flows through a conductor is directly proportional to the voltage across the conductor; that is, $\dfrac{V}{R}$ = constant; the constant is termed the resistance of the conductor
- **ohmic device** – a component with constant resistance; i.e. a device that exhibits a proportional relationship between current and voltage: $R = \dfrac{V}{I}$

_____

_____

_____

_____

_____

_____

_____

_____

_____

2 Draw a flow chart or mind map to relate and link the following key terms.

- **conductor** – a material of low resistance that allows the flow of electrons, e.g. metals

- **insulator** – a material that inhibits the flow of electrons, e.g. rubber

- **non-ohmic device** – a component that does not provide a constant resistance: $R \neq \dfrac{V}{I}$

- **Ohm's law** – the current that flows through a conductor is directly proportional to the voltage across the conductor; that is, $\dfrac{V}{R} =$ constant, which is termed the resistance of the conductor

- **ohmic device** – a component with constant resistance; i.e. a device that exhibits a proportional relationship between current and voltage: $R = \dfrac{V}{I}$

- **semiconductor** – a material that conducts electricity less readily than a conductor but more readily than an insulator; its resistance varies with temperature

- **resistance** – a quantity representing the opposition to the flow of electrical charge throughout a given material. Resistance is measured in ohms, Ω, and is the ratio between potential difference and current

- **resistivity, ρ** – a measure of how much a material opposes the flow of electrical charge; it has the unit $\Omega\,$m

10.2 Resistance

An investigation was conducted to measure the values of the resistances of two different types of light globes, represented by globe A and globe B. The results were plotted on a set of axes, as shown in Figure 10.2.1.

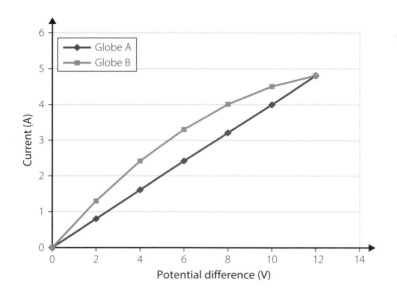

FIGURE 10.2.1 Current versus potential difference for globe A and globe B

1 Draw the circuit diagram required to obtain this data.

2 Which globe is behaving as an ohmic resistor? Explain your reasoning.

_____

_____

3 Suggest possible reasons why the other globe is not behaving as an ohmic resistor.

_____

_____

9780170412551

4 a Calculate the resistance of globe A to one decimal place using Ohm's law, $R = \dfrac{V}{I}$.

_____

_____

_____

_____

_____

_____

_____

b What is happening to the value of the resistance of globe B as the potential difference across it increases?

_____

_____

5 What safety precautions need to be implemented to address the risks of performing this investigation?

_____

_____

6 Suggest why many types of lights, such as LED downlights, are designed to operate from a 12-volt transformer, rather than directly from the 240 V supply.

_____

_____

10.3 | Conductors and insulators

1 Draw a diagram to illustrate the behaviour, location and motion of electrons in a conductor.

2 List several examples of insulators, conductors and semiconductors in Table 10.3.1.

TABLE 10.3.1 Examples of insulators, conductors and semiconductors

INSULATORS	CONDUCTORS	SEMICONDUCTORS

3 Describe the property of metals that enables electrons to flow readily.

_____

_____

_____

_____

10.4 Resistance and resistivity

1 State Ohm's law both in written and mathematical form.

_____

_____

_____

_____

2 Name the three factors that can vary the resistance of a conductor.

_____

_____

_____

3 State the formula for resistivity, ρ.

_____

_____

4 Explain the effect of temperature on the resistance of a conductor.

_____

_____

9780170412551

5 Graph the current and voltage values provided in Table 10.4.1 and determine if the resistor is ohmic or non-ohmic.

TABLE 10.4.1 Determining whether a device is ohmic or non-ohmic

CURRENT (A)	VOLTAGE (V)
0.00	0.0
0.19	0.3
0.40	0.6
0.59	0.9
0.78	1.2
1.02	1.5
1.21	1.8

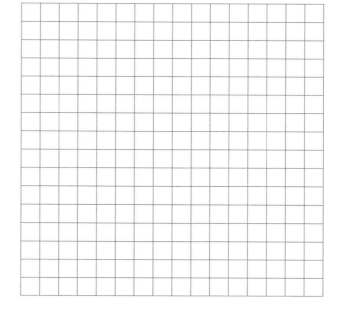

10.5 | Systematic and random errors

1 Provide two examples of systematic experimental errors.

_____

_____

2 Provide two examples of random experimental errors.

_____

_____

3 Explain the difference between systematic and random errors.

_____

_____

_____

_____

4 Explain why systematic errors are also often called 'one-sided' errors.

_____

_____

10.6 | The invention of the transistor

The development of semiconductors, from the first germanium 'crystal set' radio to the modern microprocessor containing millions of transistor connections, was largely driven by the need for reliable, portable radio transceivers. Parallel to this need was the development of more sophisticated and complex electronic circuits that were based on an increasing number of valves — unreliable, power-hungry and bulky vacuum tubes.

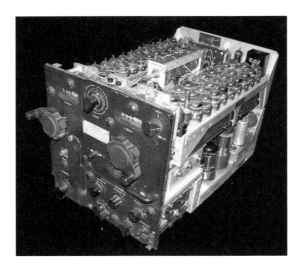

FIGURE 10.6.1 Device from an old radio navigation unit

Figure 10.6.1 shows such a device from an old radio navigation unit, the modern equivalent of which could easily fit in the palm of your hand. The increased knowledge of the behaviour of semiconducting elements — germanium, and later silicon — along with the process of 'doping', allowed huge improvements in electronics to occur, improvements that today we largely take for granted, but without which our modern society could not function.

When metals are heated to very high temperatures, free electrons are emitted from the surface of the metal. This is known as thermionic emission. Thermionic devices, therefore, require a large amount of external energy in the form of heat to operate. A solid state device uses semiconductors to direct the flow of electrons and does not require a heating circuit.

Some advantages of solid state devices over thermionic devices are listed below.

▶ *Miniature size:* The relatively large size of thermionic devices limits their uses. Solid state devices such as diodes and transistors are considerably smaller (in millimetres). Further reduction in size can be achieved in a microchip, which may contain millions of transistors in an area the size of a fingernail. The trend to miniaturisation of electronic devices, such as mobile phones, tablets and laptop computers, means the tiny solid state devices are much preferred.

▶ *Durable and long lasting:* Solid state devices are quite tough and can withstand a reasonable amount of physical impact. (Dropping a transistor might not necessarily break it.) Thermionic devices are made from glass bulbs, which make them extremely fragile, so they need to be handled with care. Also, solid state devices generally have a longer life span than thermionic devices, which must be replaced after a number of uses.

▶ *Greater operational speed:* Solid state devices operate at a much faster rate than thermionic devices. This makes them particularly valuable in the production of fast-operating microchips and microprocessors. In addition, solid state devices will function as soon as they are switched on, while thermionic devices require warming up.

▶ *More energy efficient:* Thermionic devices require very high voltages for their operation; solid state devices can function at voltages of less than 1 volt. In addition, a large amount of heat is dissipated during the operation of thermionic devices so that a considerable amount of energy is wasted. Solid state devices, however, only dissipate a small amount of energy during their operation.

▶ *Cheap to produce:* Solid state devices are much cheaper to make than the thermionic devices, so they are more economical when large numbers are needed.

Use the information above, as well as performing further research, to respond to the following questions.

9780170412551

QUESTIONS

1 What were 'valves' made from?

_____

_____

2 What were the reasons for the intense research to develop the transistor?

_____

_____

_____

_____

3 List the advantages of transistors over devices based on valves.

_____

_____

_____

4 List two ways in which the invention of transistors had a positive impact on society.

_____

_____

_____

1 Electrical resistance may be best defined as:

   A a measure of efficiency of electric current.

   B a measure of the opposition of a material to the flow of electrons.

   C a force that opposes motion.

   D a unique value, measured in $\Omega\,m$, expressing opposition to the flow of current.

2 What can a current of 20 mA also be written as?

   A 0.002 A

   B 0.2 A

   C 20 000 A

   D 0.020 A

3 An electric heater is rated at 1200 W and is connected to a 240 V supply. The current that the heater draws is equivalent to:

   A 5 A.

   B 0.5 A.

   C 0.2 A.

   D 2 A.

4 A series circuit was constructed using a 6-volt battery, and two resistors of $R_1 = 15\,\Omega$, and $R_2 = 9\,\Omega$.

   a Draw the series circuit diagram.

   b Calculate the current flowing through each resistor.

c Find the potential difference across each resistor.

_____

_____

d Draw an equivalent, simpler circuit.

```

```

5 In what fields of employment would a knowledge of electric circuits and/or electronics be useful?

_____

_____

6 State the formula that relates resistance to resistivity, cross-sectional area and length.

_____

_____

7 Explain why, when an extra resistor is added in parallel to existing resistors, the current in the circuit increases rather than decreases.

_____

_____

_____

8 A circuit comprises a 9.0 V battery, two 8.0 Ω resistors in parallel and two 7.0 Ω resistors in series with each other and with the parallel resistors.

a Draw this circuit.

```

```

b Find the total current flowing around the circuit.

_____

_____

c Find the potential difference across each resistor.

_____

_____

d Find the total power of this circuit.

_____

_____

9 Using three 2Ω resistors and two 5Ω resistors, construct a circuit with a total resistance of 5.5Ω.

9780170412551

11 Circuit analysis and design

LEARNING

Summary

- The application of electricity in electronic circuits has made devices faster, more powerful and significantly smaller.

- When electric charges run through appliances, they transform energy into various forms such as light, heat and radiant sound energy.

- In household circuits, lights, power points and appliances are connected in a variety of arrangements, frequently in parallel. Circuit breakers are included to prevent damage due to an oversupply of current.

- Power is a measure of the energy difference per unit of time. Since $P = V \times I$ and $V = I \times R$, then $P = I^2 \times R$ and $P = \dfrac{V^2}{R}$.

- In series circuits:
  - electrons have one pathway along which to travel. The current is the same throughout the circuit, while the potential difference is shared between each of the circuit elements
  - the potential difference is shared: $V_T = V_1 + V_2$
  - there are no junctions, so the current in each resistor is the same: $I_T = I_1 = I_2$
  - the equivalent resistance is given by $R_T = R_1 + R_2$.

- Any series circuit can be reduced to a single source and a single equivalent resistor.

- In parallel circuits:
  - electrons have two or more pathways along which to travel. The total number of charged particles that arrive and leave a junction each second is the same. The potential difference is the same across parallel circuit elements
  - the current is shared between circuit elements
  - the potential difference is the same across each resistor: $V_T = V_1 = V_2$
  - the total current in the circuit is shared between the resistors: $I_T = I_1 + I_2$.

- In a parallel circuit the equivalent resistance is given by: $\dfrac{1}{R_T} = \dfrac{1}{R_1} + \dfrac{1}{R_2}$.

- Electric circuits can be reduced to a simple arrangement of one equivalent source and one equivalent load.

- Ohm's law and Kirchhoff's current and voltage laws may be applied to analyse circuit components.

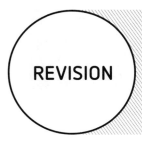

11.1 | Circuit analysis and design key terms

1 Demonstrate your understanding of the flow of charge and energy throughout an electric circuit by writing a paragraph to relate the key terms below. Use circuit diagrams where necessary.

- **combination circuit** – circuits that contain both series and parallel components

- **current, I** – the rate of flow of charge, that is, charge per unit time, measured in amperes, A: $I = \dfrac{q}{t}$

- **Kirchhoff's current law (first law)** – the total current arriving at a junction within an electrical circuit is equal to the total current leaving the junction

- **Kirchhoff's voltage law (second law)** – for any closed loop in an electric circuit, the sum of the potential differences must be zero

- **potential difference, V** – potential energy per unit charge, voltage: $V = \dfrac{W}{q}$

9780170412551

2   Draw a flow chart or mind map to relate and link the following key terms.

- **combination circuit** – circuits that contain both series and parallel components
- **current, *I*** – the rate of flow of charge, that is, charge per unit time, measured in amperes, A: $I = \dfrac{q}{t}$
- **emf** – electromotive force; source of potential energy per unit charge
- **parallel circuit** – a circuit with multiple paths through which current can flow
- **series circuit** – a circuit with only one path through which the charge can flow

11.2 Drawing electric circuits

1 Draw a simple series circuit with a single resistor and a battery of cells.

2 Draw a simple electric circuit with a battery of cells and two resistors connected in parallel.

3 Contrast a series circuit with a parallel circuit.

_____

_____

_____

4 Design a circuit that uses a 12V battery as the only source of charge, two $10\,\Omega$ resistors connected in parallel, which have a current of 1.0A flowing through each, and one other resistor of $x\,\Omega$ connected in series.

  a Draw the circuit using the correct symbols.

  b Determine $x\,\Omega$, the value of the third resistor.

_____

_____

_____

  c Given that the battery can delivery 2.8×10^{23} C of charge before being depleted, calculate for how long the circuit can operate. Recall that 1.0 A represents 6.25×10^{18} C s^{-1}.

_____

_____

_____

_____

_____

9780170412551

11.3 | Safety devices in circuits

1 Research the role of the Electrical Regulatory Authorities Council (ERAC) and write a brief outline of its main responsibilities.

_____

_____

2 In many parts of the world including Australia, power points have three-pin connections, allowing for active, neutral and earth wires. Older circuits have colour-coded wires as active (red), neutral (black) and earth (green). These were changed to active (brown), neutral (blue) and earth (helical green and yellow stripes). Why was this colour change made? _Note:_ For three-phase systems there is a different colour for each phase.

_____

_____

_____

3 The earth wire is connected to the neutral at the fuse box, and is grounded through a good conductor. What is the purpose of having an earth connection from the power point to an appliance?

_____

_____

_____

4 The diameter of wiring used for lighting circuits is 2.5 mm, while the diameter of stove wiring is 6.0 mm. What does this imply about the difference in the current in these circuits?

_____

5 What would be the implications on safety of using wiring for lighting circuits with a diameter of 6.0 mm instead of 1.0 mm?

_____

_____

11.4 | Light-emitting diodes

Light-emitting diode (LED) lighting has become the preferred source of lighting in housing, commercial and public spaces in recent years. Many televisions, smart phones and laptop computers use LED technology for fast response times and crisp, clear images.

QUESTIONS

1 Investigate the future uses of LEDs and find out why they are preferred over other lighting sources. Use a range of websites, such as the Edison Tech Center website, to determine how LEDs work. Select two other forms of lighting to research.

Compare the forms of light sources, including how they work, their typical application and their efficiency.

2 List a total of eight different types of light sources in Table 11.4.1. Classify them into what you would consider 'high-tech' (electronics based) versus 'low-tech' (of a simple design).

TABLE 11.4.1 Comparison of light sources

LIGHT SOURCE	OPERATION	TYPICAL APPLICATION	APPROXIMATE EFFICIENCY
LED			
Alternate light source 1			
Alternate light source 2			

11.5 | Battery storage systems for households

One of the limitations of photovoltaic (PV) solar panels used by individual households is their inability to produce electrical energy at night. At night, to use the energy produced during the day, households need to install a bank of batteries: lead-acid batteries (similar to the batteries found in cars) or lithium ion batteries (such as used in the Powerwall system).

The storage capacity of such 12 volt batteries is measured in amp-hours. A battery with a storage capacity of 100 amp-hours (Ah) can theoretically deliver a current of 1 amp for 100 hours at 12 volts before losing all of its charge. Further, an inverter must be used to convert the direct current (DC) into alternating current (AC), so that it is compatible with the mains power supply to allow appliances in the house to operate normally.

In this exercise, we will assume that the household in question uses an average of 2.0 kW of electrical power every evening for 6 hours.

QUESTIONS

1 How many kilowatt-hours (kWh) of electrical energy will need to be stored in the batteries?

_____

_____

2 How many 12-volt batteries would need to be connected together in series to deliver 240 V?

_____

_____

3 What current is required to deliver 2.0 kW of power at 240 V?

_____

_____

4 Given the current required, calculate the minimum amp-hour capacity of the batteries used in a battery storage bank that delivers 2.0 kW over the period of 6 hours.

_____

_____

9780170412551

11.6 | Electronic devices

Almost every appliance that we use in our homes has some sort of electronic circuit in its design. Only a few decades ago, electronic circuits were found only in the most expensive appliances, such as TVs and radios. Nowadays, ovens, smart phones, coffee machines, washing machines, dishwashers, cameras, medical equipment, cars, alarms and fridges all have complex electronic circuits built into their design.

QUESTIONS

Imagine that you have been transported 30 years into the future.

1 Given the rapid advances made in the application of electronic devices up until the 2010s, propose the devices and appliances you think might be controlled in the 2040s and what a typical household might be like.

_____

_____

_____

_____

_____

2 Suggest how developments in electronics might alter communication devices in the future.

_____

_____

_____

3 What impact has the miniaturisation of electronic devices had on our society since their invention in the 20th century?

_____

_____

_____

1 Power may be calculated by using a range of formulas. Which of the following is not used to calculate power?

A $P = V \times I$

B $P = V \times \dfrac{V}{R}$

C $P = I \times R$

D $P = I^2 \times R$

2 Four $20\,\Omega$ resistors are connected in parallel. What is the equivalent resistance?

A $80\,\Omega$

B $5\,\Omega$

C $20\,\Omega$

D $0.2\,\Omega$

3 A $2\,k\Omega$ resistor is placed in a circuit connected to a voltage potential of $240\,V$. What is the resulting current?

A $480\,A$

B $120\,A$

C $8.33\,A$

D $0.12\,A$

4 State the formula for determining the total resistance in a series circuit.

_____

5 State the formula for determining the total resistance in a parallel circuit.

_____

6 State Kirchhoff's current law.

_____

_____

7 State Kirchhoff's voltage law.

_____

_____

8 A circuit with $24\,V$ battery has two $5\,\Omega$ resistors connected in series. Determine the current that flows in the circuit.

_____

_____

9 A 1.2 kW toaster is connected to a 240 V mains supply. Determine the following.

a The current drawn

_____

_____

b The resistance of the toaster

_____

_____

c The power dissipated across the toaster

_____

_____

d The energy lost to heat if the toaster is used for a period of two minutes

_____

_____

10 A 400 W electric hair dryer is plugged into the 240 V mains supply.

a Calculate the amount of current drawn by the hair dryer.

_____

_____

b Calculate the resistance of the device.

_____

_____

c How much power is dissipated from the heating element of the dryer?

_____

_____

d Calculate the amount of energy lost to heat energy when the hair dryer is used for a period of 3 minutes.

_____

_____

LINEAR MOTION AND WAVES

- Topic 1: Linear motion and force
- Topic 2: Waves

9780170412551

Summary

▶ In order to measure quantities, scales are used.

▶ If only one scale, such as distance or time, is used to measure a quantity, the quantity is a scalar.

▶ When a quantity is derived only from scalar quantities, the quantity, such as speed or acceleration, is also scalar.

▶ A vector involves two or more scales: a scalar value and a second value, such as direction, from another scale.

▶ Vectors can be represented by arrows.
  - Length is proportional to magnitude.
  - Direction is from tail to head of arrow.

▶ Vectors can be multiplied by a scalar multiplier, k.
  - If $k > 1$, the vector retains its direction, but becomes larger in magnitude.
  - If $0 < k < 1$, the vector retains its direction, but becomes smaller in magnitude.
  - If $k = {}^{-}1$, the vector reverses direction.

▶ Vectors can be added and subtracted.
  - Vectors can be added head to tail: $\vec{a} + \vec{b}$.
  - Vector subtraction is the addition of the negative vector: $\vec{a} - \vec{b} = \vec{a} + ({}^{-}\vec{b})$.
  - Along a straight line, vectors add like scalars, using positive and negative number line rules.

▶ Objects move along number lines according to algebraic rules associated with positive and negative numbers.

▶ Displacement is directed distance:
$$\vec{s} = \vec{d}_2 - \vec{d}_1.$$

▶ The distance travelled is the magnitude of the displacement.
  - For any one displacement, $s = \left| \vec{d}_2 - \vec{d}_1 \right|$.
  - For multiple displacements, the distance travelled is the sum of the individual displacements.

12.1 | Scalar and vector quantities

The *Bureau International des Poids et Mesures* (BIPM) defines the *Systeme Internationale* (SI) units for scales of fundamental quantities, such as mass, length and time, as well as derived quantities.

TABLE 12.1.1 Scalar and vector quantities

QUANTITY AND USUAL SYMBOL	MEASUREMENT SYSTEM
SCALAR QUANTITIES	
mass, m	kilogram; kg
distance, d	length; m
time, t	second; s
electric current, I	ampere; A
temperature, T	kelvin; K
luminous intensity, I	candela; cd
amount of substance, n	mole; mol
energy, E	joule; J
VECTOR QUANTITIES	
displacement, $\bar{S}$	length; m and angle
velocity, $\bar{v}$	speed; $m\,s^{-1}$ and angle
acceleration, $\bar{a}$	acceleration; $m\,s^{-2}$ and angle
force, $\bar{F}$	newton; N and angle
momentum, $\bar{p}$	$kg\,m\,s^{-1}$ and angle

WORKED EXAMPLES

1   Consider the quantities of *time*, *speed*, *force*, *velocity* and *momentum*.

   a   Identify the vector quantities.

   b   Explain why the quantities are vector in nature.

2   **a**   Add 2 km north to 2×3 km north.

   b   Take 3 km north from 5 km north.

ANSWER

1   **a**   Force, velocity and momentum are vector quantities.

   b   Each requires a direction as well as the scalar quantities needed to derive them.

2   **a**   2 km north + $2 \times$ (3 km north) = 2 km north + 6 km north = 8 km north

   b   5 km north $-$ 3 km north = 5 km north + $^-$(3 km north) = 5 km north + 3 km south = 2 km north

QUESTIONS

1 Consider the quantities *displacement, temperature, speed, force* and *momentum.*

a Identify the scalar quantities.

_____

_____

b Explain why the quantities are scalar in nature.

_____

_____

_____

2 Explain why velocity is a vector quantity.

_____

_____

_____

3 Explain why *acceleration* can be both a scalar and a vector quantity.

_____

_____

_____

4 Multiply the following vectors:

a 3×6 km west

_____

b 4×12 km south

_____

c $^-1 \times 8$ km east

_____

d $^-5 \times 2$ km north-west.

_____

5 Find the following:

a 2 km south + 3 km south

_____

b 4 km west + 3×5 km west

_____

c 8 km east − 15 km west

_____

d 2 × 3 km north − ⁻4 × 6 km south.

_____

12.2 Vector representation

1 Use the same scale to represent the following vectors as arrows.

a 100 m, south

b 75 m, N45°E

c ⁻120 m, S50°W

d 180 m, 225 true

e ⁻90 m, 105° true

2 A cyclist travels 5.0 km north, turns left and travels 3.0 km before turning south for 7.0 km. Finally, the cyclist comes to a stop after a further 6.0 km to the east.

a Draw the cyclist's trip to scale.

b Find the distance travelled by the cyclist.

_____

_____

c Use a ruler and protractor to measure and report the displacement:

i in quadrant notation

_____

_____

ii in true bearing notation.

_____

_____

9780170412551

3 $\vec{a}$ *and* $\vec{b}$ are vectors such that:

$\vec{a} = 6.0\ m$, N30°E

$\vec{b} = 8.0\ m$, N60°W.

Draw scale diagrams to specify the following vectors:

a $^+2\vec{a}$

b $^-\vec{a}$

c $\vec{a} + \vec{b}$

d $2\vec{a} + 3\vec{b}$

e $\vec{a} - \vec{b}.$

9780170412551

4 The fictional town of Cartesia is mapped out in a series of north–south roads that intersect with a series of east–west roads. The centres of the intersections are 300 m apart. The tallest building, right in the middle of town, is a pizza shop called Galvani's Pizza, named after the owner. The building straddles Curie Parade and Lange Avenue, such that vehicles can travel on both roads. Galvani's delivers pizzas by motorbike, ridden by Richard, or by a drone called Lise. The motorbike can travel on the roads but the drone flies direct to the customer across the tops of the buildings.

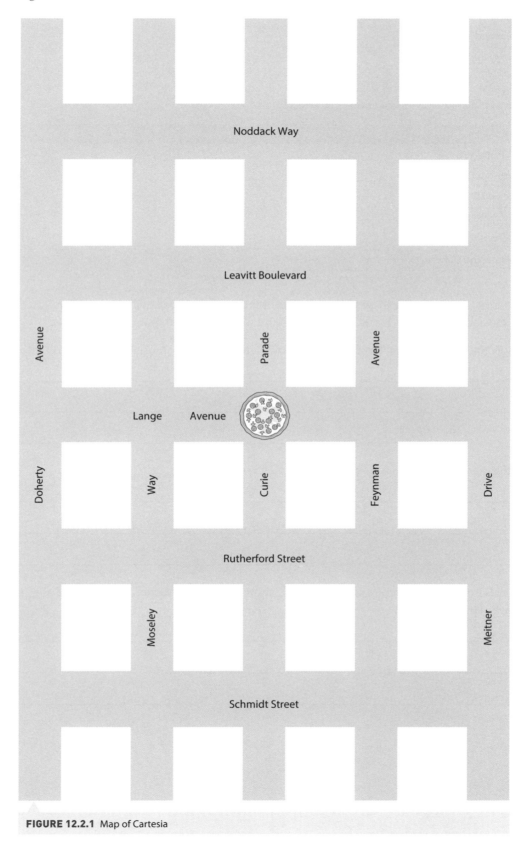

FIGURE 12.2.1 Map of Cartesia

a Complete the table to specify both distance and compass angle for the vectors from Galvani's Pizza to the following intersections in Cartesia.

INTERSECTION	DISTANCE	COMPASS ANGLE
Meitner and Schmidt		
Rutherford and Meitner		
Feynman and Noddack		
Leavitt and Moseley		
Doherty and Schmidt		

b Using east and north as positive directions, specify the easterly and northerly components relative to Galvani's Pizza of the following intersections in Cartesia.

INTERSECTION	NORTHERLY COMPONENT	EASTERLY COMPONENT
Lange and Feynman		
Lange and Doherty		
Curie and Rutherford		
Curie and Noddack		
Moseley and Schmidt		
Noddack and Doherty		

5 Lise, the drone, carries pizzas to two locations. The drone is programmed to fly to a position, 721 m, N56°E from Galvani's to deliver the first pizza. It is then to fly 900 m west to deliver the second pizza.

a Identify the position of the first delivery point by reference to the distance from the nearest intersection.

_____

_____

_____

b Specify the vector from Galvani's to the second delivery point.

_____

_____

_____

c Lise flies from the intersection of Moseley and Schmidt to the intersection of Leavitt and Meitner. Find the change of position vector.

_____

_____

_____

9780170412551

12.3 | Movement along a straight line

WORKED EXAMPLE

An object at position A at $^+8\,\text{cm}$ from the origin is then moved to position B, which is at $^+19\,\text{cm}$.

a Find the distance between A and B.

b Find the displacement of A relative to B.

c Determine the displacement of B relative to A.

ANSWER

a $s = \left| \vec{d}_2 - \vec{d}_1 \right|$

$\Rightarrow s = \left| {}^+19\,\text{cm} - {}^+8\,\text{cm} \right|$

$\Rightarrow s = 11\,\text{cm}$

b $\vec{s} = \vec{d}_2 - \vec{d}_1$

$\Rightarrow \vec{s} = {}^+19\,\text{cm} - {}^+8\,\text{cm}$

$\Rightarrow \vec{s} = {}^+11\,\text{cm}$

c $\vec{s} = \vec{d}_2 - \vec{d}_1$

$\Rightarrow \vec{s} = {}^+8\,\text{cm} - {}^+19\,\text{cm}$

$\Rightarrow \vec{s} = {}^-11\,\text{cm}$

QUESTIONS

1 Find the following displacements and the corresponding distance travelled from the origin.

a $^+4.5\,\text{cm} + {}^+6.5\,\text{cm}$

b $^-40.5\,\text{cm} - {}^-6.5\,\text{cm}$

c $^-3.5\,\text{cm} + {}^-2.8\,\text{cm} - {}^-19\,\text{mm}$

d $^+55\,\text{m} + {}^+65\,\text{cm} + {}^+95\,\text{cm} + {}^-9.2\,\text{m}$

2 Starting from home, Linh walks back and forth along a straight footpath. Linh's walk is shown in Figure 12.3.1. During segment A, Linh moves to position A. Similarly, Linh moves to positions B, C, D, F and G at the end of the corresponding sections. E is a particular point within section D. The furthest distance that Linh walks from home is 100 m.

FIGURE 12.3.1 Linh's walk

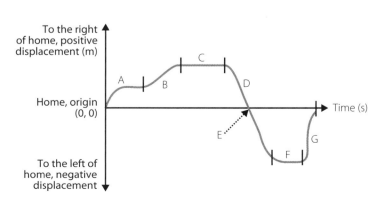

a   The sentences below describe segments of Linh's walk as shown on the graph in Figure 12.3.1. Place the letter from the graph next to the corresponding sentence.

SEGMENT	DESCRIPTION
	Linh walks past home without stopping.
	Linh goes home, walking the fastest and arriving from the left.
	Linh leaves home, turns right and walks for a while at more or less constant speed, before slowing almost to a stop.
	Linh stops for a while.
	Linh speeds up and goes further away from home, before slowing again.
	Linh nearly stops before turning around.
	Linh walks back towards home, going well past the front gate.

b   Place a scale on the displacement axis. Use the scale to identify the displacements of positions A, B, C, D, E, F and G.

c   Specify the following displacements:

i $\vec{A}$ = position of point A relative to the origin

_____

ii $\vec{B}$

_____

iii $\vec{C}$

_____

iv $\vec{C}$ relative to $\vec{D}$

_____

v $\vec{F}$ relative to $\vec{B}$

_____

vi $\vec{D}$ relative to $\vec{A}$.

_____

d   Suggest and justify a suitable value for the time at point E on the time axis.

_____

_____

_____

_____

9780170412551

1 Which of the following lists contains more than one vector quantity?

   A Distance, speed, velocity, acceleration

   B Acceleration, distance, speed, time

   C Displacement, speed, time, mass

   D Mass, force, time, speed

2 What is the difference between ⁻2.86 m and ⁻9.47 m?

   A ⁺12.3 m

   B ⁻12.3 m

   C ⁺6.61 m

   D ⁻6.61 m

3 Position B is 17 cm from position A, which is at ⁺6.5 cm. Which of the following statements best describes the position of B?

   A B is clearly at ⁺23.5 cm only.

   B B is clearly at ⁻10.5 cm only.

   C B could be at ⁺23.5 cm or ⁻10.5 cm.

   D The displacement of B relative to A is ⁺17 cm.

4 A toy car is moved along a ruler from ⁺5 cm to ⁺72 cm, then arrives at the origin after moving to ⁻29 cm. What is the displacement of the toy car, relative to the starting position and the distance moved, respectively?

   A 0 cm; 63 cm

   B ⁻98 cm; 63 cm

   C ⁻5 cm; 160 cm

   D ⁺5 cm; 160 cm

5 From which of the following is an azimuth angle measured clockwise?

   A North

   B South

   C East

   D West

6 A baseball is thrown 30 m north, then 30 m east. The respective distance travelled by the baseball and the magnitude of the displacement of the ball is nearest to:

A 60 m; 42 m.

B 42 m; 60 m.

C 60 m, 45° true; 42 m.

D 42 m, 45° true; 60 m.

7 Vector $\vec{A}$ is a displacement of 90 m 200° true. What is its displacement in quadrant-bearing notation?

A S20°W

B W20°S

C 90 m, W20°S

D 90 m, S20°W

8 A true bearing of 310° is equivalent to:

A N50°W but not W40°N.

B N40°W but not W50°N.

C N40°W and W50°N.

D N50°W and W40°N.

9 Vector R is a displacement of 20 m, 270° true. Vector T is a displacement of 20 m, 90° true. What is the vector sum, $T + {}^{-}R$?

A 0

B 40 m, N45°W

C 40 m, west

D 40 m, east

10 If $\vec{A} = {}^{-}45$ m, $120°$, then ${}^{-}2\vec{A} =$

A ${}^{+}90$ m, ${}^{-}240°$

B ${}^{-}90$ m, ${}^{-}240°$

C ${}^{+}90$ m, $120°$

D ${}^{-}90$ m, $120°$

11 Compare scalar and vector quantities, with reference to distance and displacement.

_____

_____

_____

_____

_____

9780170412551

12 An ecologist watches an ant as it moves in a straight line from a position 4.5 m from the ant nest. The ant moves a further 1.7 m away from the nest. The ant then marches past the nest until it is 1.9 m from the nest. It then returns to the nest. Show your working for the following.

a Calculate the distance travelled by the ant.

_____

_____

_____

b Calculate the final displacement of the ant relative to its initial position.

_____

_____

_____

c Determine the displacement of the ant relative to its initial position when it is at its furthest position from the nest.

_____

_____

_____

13 Vectors $\vec{A}$ and $\vec{B}$ are specified below:

$\vec{A} = 20$ m, azimuth $120°$ $\vec{B} = 15$ m, S30°W

a Use the grid provided to construct the following:

i $\vec{A}$

ii $^-\vec{B}$

iii $\vec{A} - \vec{B}$

b Use quadrant bearings to specify $\vec{A} + \vec{B}$ completely.

_____

_____

14 A fielder walks directly north of the centre of a cricket ground on a radius of 60 m. After walking for 20 m, the fielder turns west and arrives at the edge of the ground. Use a scale drawing to find the displacement of the fielder.

_____

_____

_____

9780170412551

LEARNING

Summary

▶ The SI unit for speed, velocity and acceleration are derived from their definitions:

$$[v] = \frac{[s]}{[t]} = \frac{\text{metre}}{\text{second}} = \text{m/s, m s}^{-1} \qquad [a] = \frac{[v]}{[t]} = \frac{\text{metre per second}}{\text{second}} = \text{m/s, m s}^{-2}$$

▶ Displacement (vector) is directed distance (scalar); velocity (vector) is directed speed (scalar); and acceleration is either rate of change of velocity (vector) or rate of change of speed (scalar).

▶ With movement along a straight line, intervals reduce to the difference between directed numbers.
- Distance interval: $s = |\vec{d}_2 - \vec{d}_1|$
- Displacement interval: $\vec{s} = \vec{d}_2 - \vec{d}_1$
- Time interval: $t_2 - t_1$

▶ All measures of speed and velocity are averages because they rely on distance or displacement intervals and time intervals.

▶ Average speed and average velocity
Speed:

$$v_{av} = \frac{s}{t} = \frac{|\vec{d}_2 - \vec{d}_1|}{t_2 - t_1}$$

Velocity:

$$\vec{v}_{av} = \frac{\vec{s}}{t} = \frac{\vec{d}_2 - \vec{d}_1}{t_2 - t_1}$$

▶ Converting between units:
$1\,\text{km} = 1000\,\text{m} = 10^3\,\text{m}$
$1\,\text{h} = 60\,\text{min h}^{-1} \times 60\,\text{s min}^{-1} = 3600\,\text{s h}^{-1}$
$$\rightarrow \times \frac{1000}{3600} \text{ or } \div 3.6$$

speed $(\text{km h}^{-1}) \Leftrightarrow$ speed (m s^{-1})

$$\times \frac{1000}{3600} \text{ or } \times 3.6 \leftarrow$$

▶ Graphs of s–t and v–t are frequently used to find the speed and acceleration of movements along a straight line. Graphs and algebraic formulas amount to the same thing because they both represent the same movement. Graphs have the added advantage of being able to be analysed for non-uniform motion.

- Displacement–time graphs become distance–time graphs, with positive and negative used on the number line to denote direction.
- Instantaneous speed = gradient of s–t graph at a point in time. The gradient is found by calculations of average speed as the time interval becomes smaller and smaller. As the time interval approaches zero, the average speed and the instantaneous speed become closer and closer to the same value. Thus, the instantaneous speed approaches the gradient of the s–t graph.
- If the distance interval changes by the same amount in equal time intervals the speed is constant.

$$v = \frac{s}{t} = \text{gradient of } v\text{–}t \text{ graph}$$

Where:

v = speed

s = distance change (rise of v–t graph)

t = time interval (run of v–t graph)

- When speed changes by the same amount in equal time intervals, acceleration is constant:

$$\vec{a} = \frac{\vec{v}}{t}$$

Where:

$\vec{a}$ = acceleration

$\vec{v}$ = velocity change

t = time interval

  - For speed–time graphs: acceleration = gradient v–t graph
- Acceleration–time graphs:

Where:

v = change of speed (speed interval)

t = time interval

  - If the acceleration, a, is constant: $v = at$
  - In general, speed change = area under a–t graph

9780170412551

13.1 Displacement, velocity and acceleration

WORKED EXAMPLE

It takes 30 minutes for a car to move along a straight path from $^+25\,\text{km}$ from home to a new position that is $^-30\,\text{km}$ away from home.

At the end of the trip, calculate:

a the displacement of the car relative to the origin

b the displacement of the car relative to its original position

c the average speed of the car

d the average velocity of the car.

ANSWER

a $\vec{s} = \vec{d}_2 - \vec{d}_1$

$\Rightarrow \vec{s} = {}^-30\,\text{km} - 0\,\text{km}$

$\Rightarrow \vec{s} = {}^-30\,\text{km}$

b $\vec{s} = \vec{d}_2 - \vec{d}_1$

$\Rightarrow \vec{s} = {}^-30\,\text{km} - {}^+25\,\text{km}$

$\Rightarrow \vec{s} = {}^-55\,\text{km}$

c $v_{av} = \dfrac{\left|\vec{d}_2 - \vec{d}_1\right|}{t_2 - t_1}$

$\Rightarrow v_{av} = \dfrac{\left|{}^-30\,\text{km} - {}^+25\,\text{km}\right|}{0.5\,\text{h}}$

$\Rightarrow v_{av} = \dfrac{\left|{}^-55\,\text{km}\right|}{0.5\,\text{h}}$

$\Rightarrow v_{av} = 110\ \text{km h}^{-1}$

d $\vec{v}_{av} = \dfrac{\left|\vec{d}_2 - \vec{d}_1\right|}{t_2 - t_1}$

$\Rightarrow \vec{v}_{av} = \dfrac{{}^-30\,\text{km} - {}^+25\,\text{km}}{0.5\,\text{h}}$

$\Rightarrow \vec{v}_{av} = \dfrac{{}^-55\,\text{km}}{0.5\,\text{h}}$

$\Rightarrow \vec{v}_{av} = {}^-110\ \text{km h}^{-1}$

QUESTIONS

1 Explain the difference between speed and velocity.

_____

_____

_____

_____

_____

2 Calculate the average speed and the average velocity for the following movements along a number line. Give all answers in $m\,s^{-1}$.

a $^-6.6\,m$ to $^+9.5\,m$ in $25\,s$

_____

_____

b $^-85\,km$ to $^-95\,km$ in $125\,s$

_____

_____

c $^+68\,km$ to $^+95\,km$ in $12\,min$

_____

_____

d $^-1285\,m$ to $^+1.0\,km$ in $1.2\,s$

_____

_____

e $^+855\,km$ to $^+459\,km$ in $1.3\,h$

_____

_____

3 A runner completes one circuit of a 400 m track in 56 s. Compare the average velocity and the average speed for the runner.

_____

_____

_____

_____

_____

4 For the flying time of an aeroplane that travels from Brisbane and returns via Cairns (distance = 1395 km; flying time = 2.4 h), Cairns to Charleville (657 km; 1.5 h) and Charleville to Brisbane (464 km; 0.85 h), find:

a average speed

_____

_____

_____

_____

_____

b average velocity.

_____

_____

_____

_____

_____

13.2 | Speed and velocity

The SI units for speed, velocity and acceleration are derived from their definitions.

$$[v] = \frac{[s]}{[t]} = \frac{metre}{second} = m/s,\ m\,s^{-1} \qquad [a] = \frac{[v]}{[t]} = \frac{metre\ per\ second}{second} = m/s/s,\ m/s^2,\ m\,s^{-2}$$

WORKED EXAMPLES

1 Convert $75\,km\,h^{-1}$ to units of $m\,s^{-1}$.

ANSWER

$speed = 75\,km\,h^{-1}$

$\Rightarrow speed = 75\dfrac{km}{h}$

$\Rightarrow speed = 75 \times \dfrac{km \times 10^3\,m\,km^{-1}}{h \times 60\,min\,h^{-1} \times 60\,s\,min^{-1}}$

$\Rightarrow speed = 75 \times \dfrac{10^3\,m}{h \times 60\,min\,h^{-1} \times 60\,s\,min^{-1}}$

$\Rightarrow speed = 75 \times \dfrac{10^3\,m}{3.6 \times 10^3\,s}$ OR $\qquad \Rightarrow speed = 75\,km\,h^{-1}$

$\Rightarrow speed = \dfrac{75 \times m}{3.6 \times s}$ $\qquad\qquad\qquad \Rightarrow speed = 75\dfrac{km}{h}$

$\Rightarrow speed = \dfrac{75}{3.6} \times \dfrac{m}{s}$ $\qquad\qquad\qquad \Rightarrow speed = \dfrac{75}{3.6} \times \dfrac{m}{s}$

$\Rightarrow speed = 21\,m\,s^{-1}$ $\qquad\qquad\qquad \Rightarrow speed = 21\,m\,s^{-1}$

2 Convert $30\,m\,s^{-1}$ to $km\,h^{-1}$.

ANSWER

$speed = 30\,m\,s^{-1}$

$\Rightarrow speed = 30\dfrac{m}{s}$

$\Rightarrow speed = 30 \times 3.6\dfrac{km}{h}$

$\Rightarrow speed = 108\,km\,h^{-1}$

QUESTIONS

1 Convert the following from $m\,s^{-1}$ to $km\,h^{-1}$.

a $16\,m\,s^{-1}$

_____

_____

b $4.0\,\mathrm{m\,s^{-1}}$

_____

_____

c $24.0\,\mathrm{m\,s^{-1}}$

_____

_____

d $20.6\,\mathrm{m\,s^{-1}}$

_____

_____

e $14.9\,\mathrm{m\,s^{-1}}$

_____

_____

2 Convert the following from $\mathrm{km\,h^{-1}}$ to $\mathrm{m\,s^{-1}}$.

a $40\,\mathrm{km\,h^{-1}}$

_____

_____

b $80\,\mathrm{km\,h^{-1}}$

_____

_____

c $120.0\,\mathrm{km\,h^{-1}}$

_____

_____

d $31.5\,\mathrm{km\,h^{-1}}$

_____

_____

e $354\,\mathrm{km\,h^{-1}}$

_____

_____

3 The speed of Earth on its orbit around the Sun is $30\,\mathrm{km\,s^{-1}}$. Convert this to $\mathrm{km\,h^{-1}}$. Give the answer in scientific notation.

_____

_____

_____

13.3 | Interpreting graphs: linear motion

WORKED EXAMPLE

The graph shows the distance of a runner from the start. Find the average speed of the runner.

FIGURE 13.3.1

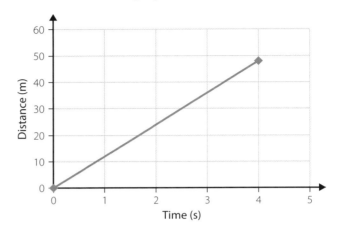

ANSWER

$$v_{av} = \text{gradient} = \frac{48\,\text{m} - 0\,\text{m}}{4.0\,\text{s} - 0\,\text{s}}$$

$$\Rightarrow v_{av} = 12\,\text{m s}^{-1}$$

SPEED–TIME GRAPHS

▶ If the speed is constant, $s = vt$.

▶ In general, distance travelled = area under v–t graph.

WORKED EXAMPLE

Figure 13.3.2 shows the speed of a car as a function of time. Calculate the distance travelled by the car between 2.0 s and 20 s.

FIGURE 13.3.2

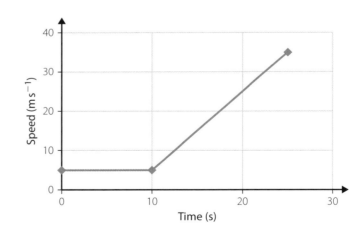

ANSWER

$s = \text{area under } v - t \text{ graph} = \text{area rectangle } (t = 2.0\,\text{s} - 10\,\text{s}) + \text{area trapezium } (t = 10\,\text{s} - 25\,\text{s})$

$$\Rightarrow s = 5.0\,\text{m s}^{-1} \times (10\,\text{s} - 2.0\,\text{s}) + \frac{1}{2}(5.0\,\text{m s}^{-1} + 25.0\,\text{m s}^{-1}) \times (25\,\text{s} - 10\,\text{s})$$

$$\Rightarrow s = 40\,\text{m} + 225\,\text{m}$$

$$\Rightarrow s = 265\,\text{m}$$

9780170412551

QUESTIONS

1 Figure 13.3.3 shows the distance travelled by a car over a 100 s period.

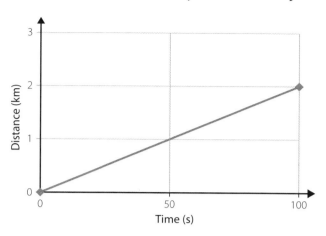

FIGURE 13.3.3

a Calculate the average speed in the first 100 s.

_____

_____

_____

b Find the instantaneous speed at 30 s.

_____

_____

_____

2 Figure 13.3.4 shows the speed of a 5000 m runner for a short burst.

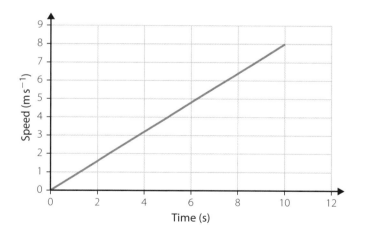

FIGURE 13.3.4

Find the distance travelled by the runner.

_____

_____

_____

3 Figure 13.3.5 shows the motion of a car over a 50 s period.

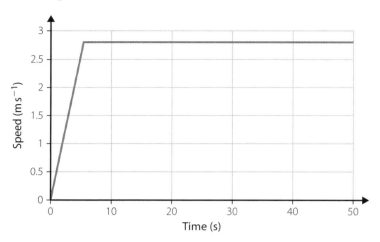

FIGURE 13.3.5

a Calculate the distance travelled in the following time intervals:

i 0–20 s

_____

_____

ii 10–30 s

_____

_____

iii 20–50 s

_____

_____

iv 4.0–38 s

_____

_____

b Calculate the average speed in the following time intervals:

i 0.0–4.0 s

_____

_____

ii 0–20 s

_____

_____

iii 4.0–40 s

_____

_____

9780170412551

4 A car travels at $45\,\text{km h}^{-1}$ for $7.0\,\text{s}$ before accelerating to $100\,\text{km h}^{-1}$ in $4.0\,\text{s}$. Sketch a graph of this motion to calculate:

a the total distance travelled

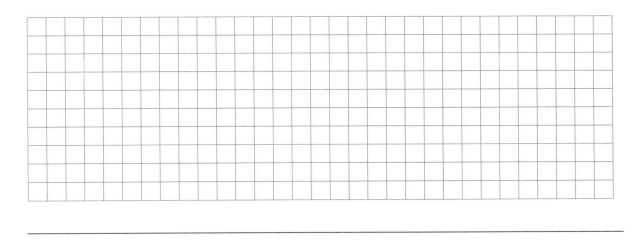

_____

_____

_____

b the average speed during the first 10 s.

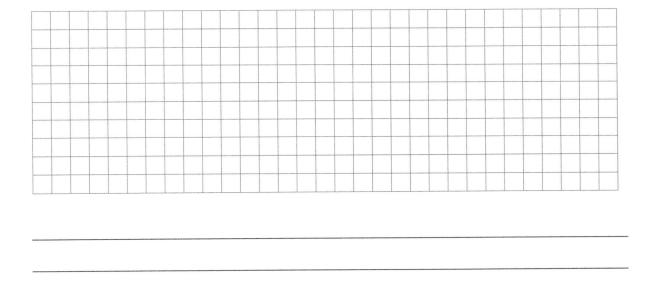

_____

_____

_____

13.4 | Acceleration

TYPE OF GRAPH	GRADIENT REPRESENTS	AREA REPRESENTS
displacement–time	speed; velocity	(–)
velocity–time	acceleration	displacement; distance
acceleration–time	(–)	change of velocity; change of speed

WORKED EXAMPLES

1 Figure 13.4.1 shows the acceleration of a sprinter during a 100 m race.

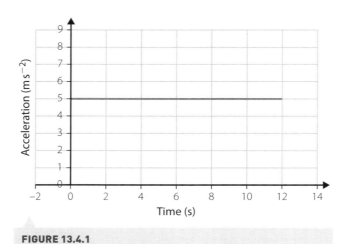

FIGURE 13.4.1

Find the speed of the sprinter after the first 2.0 s.

ANSWER

area = change of speed

$$\Rightarrow v = at$$

$$\Rightarrow v_{2s} - v_{0s} = 5.0 \text{ m s}^{-2} \times (2.0 \text{ s} - 0.0 \text{ s})$$

$$\Rightarrow v_{2s} = 10 \text{ m s}^{-1}$$

2 Rachel accelerates uniformly from stationary to a constant speed of 10 km h^{-1} in 5.0 seconds. She maintains this speed for the next 2.0 minutes.

   a Express Rachel's jogging speed in standard SI units. Show your working.

   b Sketch a speed–time graph of Rachel's run.

   c Calculate Rachel's acceleration in the first 5.0 s.

   d Calculate the distance travelled by Rachel in:

      i 5.0 s

      ii 2.0 minutes.

   e Find Rachel's average speed for:

      i the first 5.0 s

      ii the first 100 s

      iii between 2.0 s and 8.0 s.

9780170412551

ANSWER

a   speed $= 10 \text{ km h}^{-1}$

$\Rightarrow \text{speed} = 10\dfrac{\text{km}}{\text{h}}$

$\Rightarrow \text{speed} = \dfrac{10 \times \text{m}}{3.6 \times \text{s}}$

$\Rightarrow \text{speed} = \dfrac{10}{3.6} \times \dfrac{\text{m}}{\text{s}}$

$\Rightarrow \text{speed} = 2.8 \text{ m s}^{-1}$

b

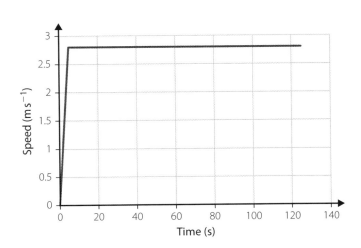

FIGURE 13.4.2

c   gradient $v - t$ graph $= a = \dfrac{v}{t}$

$\Rightarrow a = \dfrac{2.8 \text{ m s}^{-1} - 0.0 \text{ m s}^{-1}}{5.0 \text{ s} - 0.0 \text{ s}}$

$\Rightarrow a = 0.56 \text{ m s}^{-2}$

d **i**   area $= s = \dfrac{1}{2}vt$

$\Rightarrow s = \dfrac{1}{2} \times 2.8 \text{ m s}^{-1} \times 5.0 \text{ s}$

$\Rightarrow s = 7.0 \text{ m}$

  ii   area $= d = \dfrac{1}{2}vt_1 + vt_2$

$\Rightarrow s = \dfrac{1}{2} \times 2.8 \text{ m s}^{-1} \times 5.0 \text{ s} + 2.8 \text{ m s}^{-1} \times (120 \text{ s} - 5.0 \text{ s})$

$\Rightarrow s = 7.0 \text{ m} + 322 \text{ m}$

$\Rightarrow s = 329 \text{ m}$

e **i** $\text{area} = s = \dfrac{1}{2}(2.8 \text{ m s}^{-1}) \times (5.0 \text{ s} - 0 \text{ s})$

$\Rightarrow s = 7.0 \text{ m}$

$\Rightarrow v_{av} = \dfrac{s}{t} = \dfrac{7.0 \text{ m}}{5.0 \text{ s}} = 1.4 \text{ m s}^{-1}$

ii $\text{area} = s = \dfrac{1}{2}(2.8 \text{ m s}^{-1}) \times (5.0 \text{ s} - 0 \text{ s}) + (2.8 \text{ m s}^{-1}) \times (100.0 \text{ s} - 5.0 \text{ s})$

$\Rightarrow s = 7.0 \text{ m} + 266 \text{ m} = 273 \text{ m}$

$\Rightarrow v_{av} = \dfrac{s}{t} = \dfrac{273 \text{ m}}{100 \text{ s}} = 2.73 \text{ m s}^{-1}$

iii $v_{2.0 \text{ s}} = 0.56 \text{ m s}^{-2} \times (2.0 \text{ s} - 0 \text{ s}) = 1.12 \text{ m s}^{-1}$

$\text{area} = s = \dfrac{1}{2}(2.8 \text{ m s}^{-1} + 1.12 \text{ m s}^{-1}) \times (5.0 \text{ s} - 2.0 \text{ s}) + (2.8 \text{ m s}^{-1}) \times (8.0 \text{ s} - 5.0 \text{ s})$

$\Rightarrow s = 5.88 \text{ m} + 8.4 \text{ m} = 14.28 \text{ m}$

$\Rightarrow v_{av} = \dfrac{s}{t} = \dfrac{14.28 \text{ m}}{(8.0 \text{ s} - 2.0 \text{ s})} = 2.4 \text{ m s}^{-1}$

QUESTIONS

1 A car travels at 20 m s^{-1} for 4.0 minutes.

a Find the distance travelled in 2.0 s.

_____

_____

_____

b Complete the following data table to plot the distance–time graph:

TIME (s)	0	30	60	90	120	150	180	210	240
DISTANCE (m)									

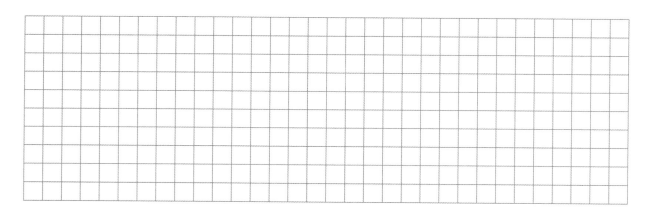

9780170412551

2 **a** Use speed–time graph (Figure 13.4.3) to deduce the:

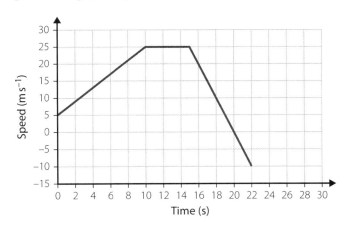

FIGURE 13.4.3

i distance–time graph

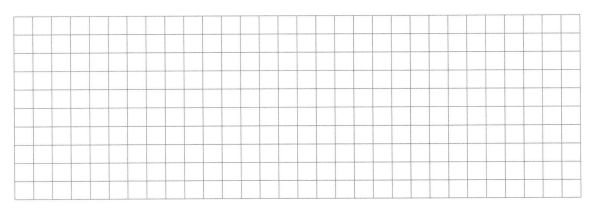

ii acceleration–time graph.

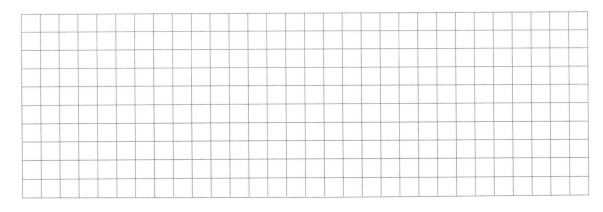

b Use the speed–time graph in Figure 13.4.3 to calculate the average speed over the following time intervals:

i 4.0 s to 10.0 s

_____

_____

_____

ii 0.0 s to 14 s

_____

_____

_____

iii 10 s to 22 s.

_____

_____

_____

3 Figure 13.4.4 shows the motion of a car in a 30 s period.

FIGURE 13.4.4

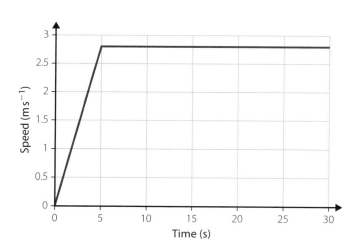

a i Calculate the acceleration in the first 15 s.

_____

_____

_____

ii Calculate the average acceleration in the first 30 s.

_____

_____

_____

iii Find the distance travelled between 10 s and 25 s.

_____

_____

_____

9780170412551

iv Calculate the average speed between 5.0 s and 25 s.

_____

_____

_____

b Draw graphs of:

i distance–time

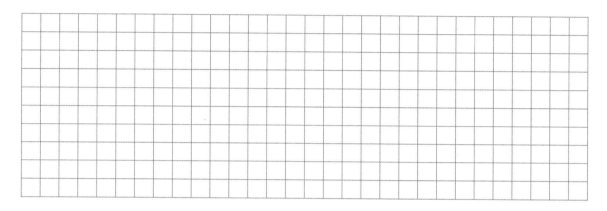

ii acceleration–time.

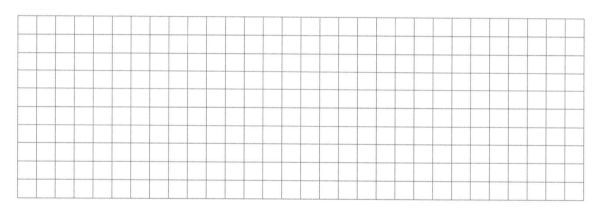

4 Figure 13.4.5 shows the acceleration of a car over a 12 s interval. The car was travelling at $15\,\text{m s}^{-1}$ before accelerating.

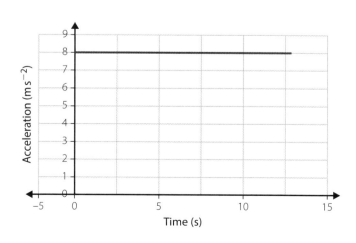

FIGURE 13.4.5

Show how to use the graph to find:

a the speed of the car after 12 s

_____

_____

b the speed gain between 4.0 s and 6.2 s.

_____

_____

_____

13.5 | Solving problems using algebra

Algebraic analysis of motion under constant acceleration makes use of the *suvat* equations, for which the following symbols apply:

- distance interval = s (s is not a particular point on a line)
- initial speed = u
- final speed = v
- acceleration = a
- time interval = t (t is not an instantaneous time)

The equations for straight-line motion with constant acceleration are:

$$v = u + at$$

$$s = ut + \frac{1}{2}at^2 \qquad s = vt - \frac{1}{2}at^2$$

$$s = \frac{u + v}{2}t$$

$$v^2 = u^2 + 2as$$

Each equation connects four variables:

- Knowing three variables, a fourth can be deduced
- With four known variables, the fifth can be deduced.

WORKED EXAMPLE

An arrow is launched vertically with a speed of $40\,\mathrm{m\,s^{-1}}$.

a Calculate the time taken to reach the highest point.

b Determine the maximum height to which the arrow rises.
  - Take air friction to be negligible.
  - Assume acceleration due to gravity = $9.80\,\mathrm{m\,s^{-2}}$.

9780170412551

ANSWER

a The acceleration due to gravity is uniform (constant) and negative with respect to the change of speed as the arrow rises:

$s = ?$ $u = 40 \, \text{m s}^{-1}$ $v = 0 \, \text{m s}^{-1}$ $a = {}^{-}9.80 \, \text{m s}^{-2}$ $t = 4.1 \, \text{s}$

$v = u + at$

$\Rightarrow t = \dfrac{v - u}{a}$

$\Rightarrow t = \dfrac{0 \, \text{m s}^{-1} - 40 \, \text{m s}^{-1}}{{}^{-}9.80 \, \text{m s}^{-2}}$

$\Rightarrow t = \dfrac{{}^{-}40 \, \text{m s}^{-1}}{{}^{-}9.80 \, \text{m s}^{-2}}$

$\Rightarrow t = 4.1 \, \text{s} \, (4.08 \, \text{s})$

b $s = ?$ $u = 40 \, \text{m s}^{-1}$ $v = 0 \, \text{m s}^{-1}$ $a = {}^{-}9.80 \, \text{m s}^{-2}$ $t = 4.08 \, \text{s}$

$s = ut + \dfrac{1}{2}at^2$

$\Rightarrow s = 40 \, \text{m s}^{-1} \times 4.08 \, \text{s} + \dfrac{1}{2}({}^{-}9.80 \, \text{m s}^{-2}) \times (4.08 \, \text{s})^2$

$\Rightarrow s = 82 \, \text{m} \, (81.6 \, \text{m})$

Alternative 1: Note that $v = 0 \, \text{m s}^{-1}$ at the top simplifies the calculation:

$s = ?$ $u = 40 \, \text{m s}^{-1}$ $v = 0 \, \text{m s}^{-1}$ $a = {}^{-}9.80 \, \text{m s}^{-2}$ $t = 4.08 \, \text{s}$

$s = vt - \dfrac{1}{2}at^2$

$\Rightarrow s = 0 \, \text{m s}^{-2} - \dfrac{1}{2}({}^{-}9.80 \, \text{m s}^{-2}) \times (4.08 \, \text{s})^2$

$\Rightarrow s = 82 \, \text{m} \, (81.6 \, \text{m})$

Alternative 2:

$s = ?$ $u = 40 \, \text{m s}^{-1}$ $v = 0 \, \text{m s}^{-1}$ $a = {}^{-}9.80 \, \text{m s}^{-2}$ $t = 4.08 \, \text{s}$

$s = \dfrac{(u + v)}{2}t$

$\Rightarrow s = \dfrac{(40 \, \text{m s}^{-1} + 0 \, \text{m s}^{-1})}{2} \times 4.08 \, \text{s}$

$\Rightarrow s = 82 \, \text{m} \, (81.6 \, \text{m})$

Alternative 3:

$s = ?$ $u = 40 \, \text{m s}^{-1}$ $v = 0 \, \text{m s}^{-1}$ $a = {}^{-}9.80 \, \text{m s}^{-2}$ $t = 4.08 \, \text{s}$

$v^2 = u^2 + 2as$

$\Rightarrow s = \dfrac{v^2 - u^2}{2a}$

$\Rightarrow s = \dfrac{(0 \, \text{m s}^{-1})^2 - (40 \, \text{m s}^{-1})^2}{2 \times ({}^{-}9.80 \, \text{m s}^{-2})}$

$\Rightarrow s = 82 \, \text{m} \, (81.6 \, \text{m})$

Thus, once the fourth variable has been deduced (in this case, t), a variety of solution strategies can be used to deduce the fifth variable (in this case, s). With experience, it is possible to select the simplest of the possibilities more often.

QUESTIONS

1 Complete the following table:

s (m)	u (m s^{-1})	v (m s^{-1})	a (m s^{-2})	t (s)
35	10	5.0		
	15	35	6.0	
		0	$^-$10	8.0
36			10	12
96	12			5.0

2 An aeroplane travelling at $40\,\text{m s}^{-1}$ accelerates uniformly to $60\,\text{m s}^{-1}$ in 10 s.

  a Calculate the acceleration of the aeroplane.

  b Determine the speed of the aeroplane after:

    i 4.0 s

    ii 10 s.

  c Calculate the distance travelled by the aeroplane while it was accelerating.

3 A rock is dropped from a height and reaches a speed of $300\,\text{m s}^{-1}$ just before it lands. Find the height from which it was dropped.
 • Ignore any air friction.
 • Assume acceleration due to gravity = $9.80\,\text{m s}^{-2}$.

9780170412551

4 A jet takes 0.95 km to reach its take-off speed of $123\,km\,h^{-1}$.

a Calculate the time taken for the jet to reach its take-off speed.

_____

_____

_____

b Determine the acceleration of the jet.

_____

_____

_____

c There is a 2.1 km safety zone at the end of the runway. The pilot needs to abort the take-off when travelling at $84\,km\,h^{-1}$. Find the minimum average acceleration needed in order for the jet to come to a stop within the safety zone.

_____

_____

_____

5 Given the maximum deceleration of a car is $8.0\,m\,s^{-2}$ and the reaction time for the driver is 1.0 s, compare the stopping distances for the car from these initial speeds.

a $90\,km\,h^{-1}\,(25\,m\,s^{-1})$

_____

_____

b $45\,km\,h^{-1}$

_____

_____

1   The graph shows the position of a particle while moving along a straight line.

FIGURE 13.5.1

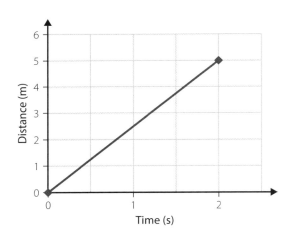

The speed of the particle at 1.5 s is closest to:

A 2.5 m s^{-1}.

B 4.0 m s^{-1}.

C 5.0 m s^{-1}.

D 10 m s^{-1}.

2   Velocity is measured as:

A distance divided by time.

B time rate of change of distance.

C time rate of change of displacement.

D the time it takes to move from one position to another.

3   It takes Van an hour to travel the first 80 km of a trip and an hour and a half for the remaining 135 km. What is Van's average speed for the journey?

A 80 km h^{-1}

B 85 km h^{-1}

C 86 km h^{-1}

D 88 km h^{-1}

4 The instantaneous speed at a particular time is deduced from a distance–time graph by finding:

    A the difference in height of the graph from start to finish and dividing by time.

    B the difference in height of the graph from start to finish.

    C the gradient of the graph at that particular time.

    D the area under the graph.

5 The graph in Figure 13.5.2 shows how the speed of a car changes over time.

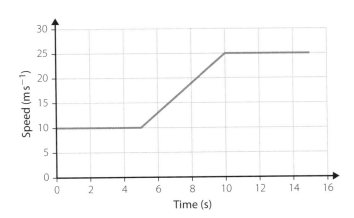 

FIGURE 13.5.2

The distance travelled by the car between 4.0 s and 10 s is:

    A 25 m.

    B 125 m.

    C 150 m.

    D 250 m.

6 The speed of a cyclist changes from $4.0\,\mathrm{m\,s^{-1}}$ to $10\,\mathrm{m\,s^{-1}}$ in 3.0 s. What is the average acceleration of the cyclist?

    A $2.0\,\mathrm{m\,s^{-2}}$.

    B $3.0\,\mathrm{m\,s^{-2}}$.

    C $6.0\,\mathrm{m\,s^{-2}}$.

    D $10\,\mathrm{m\,s^{-2}}$.

7 The area under an acceleration–time graph shows:

    A the change in acceleration.

    B the distance travelled.

    C the change in speed.

    D the average speed.

8 A car travelling at $16\,\mathrm{m\,s^{-1}}$ accelerates uniformly at $3.0\,\mathrm{m\,s^{-2}}$ for 4.0 s. The speed of the car after 4.0 s is:

    A $19\,\mathrm{m\,s^{-1}}$.

    B $20\,\mathrm{m\,s^{-1}}$.

    C $24\,\mathrm{m\,s^{-1}}$.

    D $28\,\mathrm{m\,s^{-1}}$.

9 How far will a car that accelerates uniformly from $12\,\mathrm{m\,s^{-1}}$ to $20\,\mathrm{m\,s^{-1}}$ in 4.0 s travel?

A 48 m

B 64 m

C 88 m

D 112 m

10 A car accelerates uniformly from $12\,\mathrm{m\,s^{-1}}$ to $20\,\mathrm{m\,s^{-1}}$ in 4.0 s. How far will it travel in that time?

A 48 m

B 64 m

C 88 m

D 112 m

11 A car is 65 km from a town. After 90 minutes the car is at a new position, 15 km on the other side of the town.

a Calculate the displacement of the car at the end of the trip relative to its original position.

_____

_____

_____

b Calculate the average speed of the car in $\mathrm{m\,s^{-1}}$.

_____

_____

_____

12 The speed–time graph in Figure 13.5.3 represents a cyclist's journey.

FIGURE 13.5.3

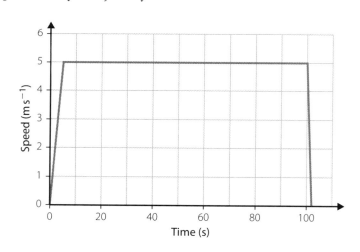

a Find the maximum speed attained by the cyclist.

_____

_____

_____

b While braking, did the cyclist decelerate at a rate greater than or less than their initial acceleration? Explain.

_____

_____

_____

c Calculate the acceleration of the cyclist in the last 2.0 s of the ride.

_____

_____

_____

d Calculate the distance travelled by the cyclist.

_____

_____

_____

e Calculate the average speed of the cyclist in the first 45 s.

_____

_____

_____

13 A ball is propelled vertically upwards from a 40 m high cliff at a speed of $8.0\,\mathrm{m\,s^{-1}}$.
- Ignore air resistance.
- Assume $g = 9.8\,\mathrm{m\,s^{-2}}$.

a Find the maximum height to which the ball rises.

_____

_____

_____

b Calculate the speed at which the ball hits the ground below the cliff.

_____

_____

_____

c Find the time that the ball spends in the air.

_____

_____

_____

14 A car is travelling at $99\,km\,h^{-1}$ when the driver notices a cow on the road. It takes $1.4\,s$ for the driver to begin braking, and a further $2.4\,s$ to stop just in front of the cow. Find the braking distance. Show all working.

_____

_____

_____

_____

_____

_____

14 Forces

Summary

▶ Forces are applied to objects and can affect their motion. Forces can cause objects to speed up, slow down, change direction or change speed and direction simultaneously.

▶ Forces are external actions applied by one object, A (*agent*), on another object, B (*receiver*). This is written in symbol form as:

  $\vec{F}$(by A on B)

▶ The interaction between objects is mutual.
  - Object A exerts a force on object B: $\vec{F}$(by A on B)
  - Object B exerts a force on A: $\vec{F}$(by B on A).

▶ Forces are vectors. They can be represented as arrows:
  - length – proportional to the magnitude of the force
  - direction – points from tail to head in the direction of the force.

▶ Like all vectors, forces can be:
  - added and subtracted geometrically
  - multiplied by a scalar, including ($^-$1), which reverses the direction.

▶ There are two types of forces:
  - contact force – force acting when two objects appear to be touching
  - non-contact force – force acting when two objects are clearly separated.

▶ Electrostatic force is the force applied by an electric charge on another electric charge:

TABLE 14.0.1 Charges of the same sign repel; charges of different signs attract

	⊕	⊖
⊕	Repel	Attract
⊖	Attract	Repel

▶ Magnetic force is the force applied by one magnetic pole on another magnetic pole:
  - same poles repel
  - opposite poles attract.

TABLE 14.0.2 Magnetic poles of the same sign repel; magnetic poles of different signs attract

	NORTH POLE	SOUTH POLE
NORTH POLE	Repel	Attract
SOUTH POLE	Attract	Repel

- Gravitational force is the force of attraction by one mass on another mass.
- Near the surface of Earth, weight is a force applied by the Earth's mass on nearby masses:
$\vec{F}(\text{by } M_E \text{ on O})$
- On Earth, weight is the gravitational force applied by the Earth's mass on another mass. Near the Earth, the gravitational force applied by the Earth on each one-kilogram of an object's mass is a constant 9.8 N.

$$\vec{g} = \frac{\vec{F}_g}{m} = 9.8 \text{ N kg}^{-1}$$

$$\Rightarrow \vec{F}_g = m\vec{g}$$

$$\text{but } \vec{F}_g = \vec{w}$$

$$\Rightarrow \vec{w} = m\vec{g}$$

- This causes a constant downward acceleration of 9.8 m s^{-2} on all masses.
- Weight scales on Earth are usually calibrated so that, when the force on a 1 kg mass is measured, the scale shows 1.0 kg, not 9.8 N.
- Scales that are calibrated in this way and taken to the Moon show the weight of a 1 kg mass is significantly less $\left(\sim \frac{1}{6} \right)$.
  - The 1 kg mass has not changed between Earth and the Moon.
  - The effect of the large, nearby mass (Earth or Moon) on the 1 kg mass is different — this is the value shown on the scale that was calibrated on Earth.
- Newton devised three laws about the action of forces, which are used to explain motion.
- **Newton's first law** – the law of inertia: every body will continue in its state of rest or of uniform motion in a straight line unless compelled by a net external force to change that state.
  - Inertia is the tendency of a body to continue in its state of rest or of uniform motion in a straight line; a kind of resistance to an applied force.
- **Newton's second law** – acceleration of a body is:
  - caused by the vector sum of forces applied to the body
  - affected by the mass of the body.

$$\vec{a} = \frac{\sum \vec{F}}{m}$$

Where: $\vec{a}$ = acceleration

$\sum \vec{F}$ = net force

m = mass

- The familiar equation for Newton's second law is found by algebraic transposition:
$$\sum \vec{F} = m\vec{a}$$
- Corollaries:

if $\sum \vec{F} = 0$, then $\vec{a} = 0$

if $\vec{a} = 0$, then $\sum \vec{F} = 0$.

- **Newton's third law:** for every action there is an equal and opposite reaction.
  - When two objects A and B interact, the action–reaction pair of forces are:
$F(\text{by A on B})$ and $F(\text{by B on A})$
  - Four conditions must be satisfied for these to be an action–reaction (or Newton 3) pair of forces. The forces must:
    - be the same fundamental type
    - be equal in magnitude
    - be opposite in direction
    - act on different objects.

9780170412551

- All surfaces apply forces to objects that are at right angles to the surface, and this perpendicular force is called the normal force (N).

- An object on a surface experiences an electrostatic force due to the surface, which is perpendicular (normal) to the surface:

  $\vec{F}_\perp$ (by surface on object)

- The surface experiences an electrostatic force due to the object:

  $\vec{F}$ (by object on surface)

- These two forces satisfy the four criteria to be an action–reaction or Newton 3 pair of forces: (i) same type, (ii) equal magnitude, (iii) opposite direction, (iv) act on different objects.

- For an object that is moving horizontally while in contact with a surface, the weight force and the normal force must be equal. This is because there is no vertical acceleration:

  $\sum \vec{F}$ (on object) $= \vec{F}_\perp$ (by surface on object) $- w = 0$

  $\vec{N} - w = 0$

  $\Rightarrow \vec{N} = w$

- The forces acting *on the same object* add to zero. Consequently, the weight force and the normal force are *not* an action–reaction pair.

- Every mass can be represented as a single, point mass.

- A free-body diagram shows all the forces applied to a single point mass.
  - Tail starts at the point of application.
  - Arrow points away from the point mass.

14.1 Forces acting on an object

WORKED EXAMPLES

1 Two cars, an Aston-Martin (A) and a Bentley (B), collide. Use the agent–receiver nomenclature to describe the forces involved in the interaction.

ANSWER

If the Aston-Martin is the agent: $\vec{F}$(by A on B)

If the Bentley is the agent: $\vec{F}$(by B on A)

2 A 120 kg payload is taken to the Moon along with the spring scales used to measure it. When weighed on the Moon with these scales, the payload appears to have shrunk to about 20 kg. Explain.

ANSWER

The weight of the payload on Earth is:

$w = mg_{\text{Earth}}$

$\Rightarrow w = 120\,\text{kg} \times 9.8\,\text{ms}^{-2}$

$\Rightarrow w = 1176\,\text{N}$

The scale shows 120 kg. The conversion factor is the gravitational acceleration, $9.8\,\text{ms}^{-2}$.
The weight of the payload on the Moon is:

$w = mg_{\text{Moon}}$

$\Rightarrow w \approx 120\,\text{kg} \times 1.7\,\text{ms}^{-2}$

$\Rightarrow w \approx 196\,\text{N}$

Using the same conversion factor, the scale shows:

$w\ (\text{scale reading}) = \dfrac{196\,\text{N}}{9.8\,\text{ms}^{-2}} = 20\,\text{kg}$

QUESTIONS

1 a Distinguish between 'contact' and 'non-contact' forces.

_____

_____

_____

b Identify three non-contact forces.

_____

_____

_____

c Explain why all forces are actually action-at-a-distance forces.

_____

_____

_____

2 Define 'agent' and 'receiver' in a two-body interaction.

_____

_____

_____

3 The north pole of magnet H is brought towards the north pole of magnet J.

a Use the symbolism, F(by ... on ...), to identify the force by H and the force by J.

_____

_____

_____

b Describe, in terms of force, the effect that H has on J and J has on H.

_____

_____

_____

4 Positively charged balloon C is brought near charged balloon D, which immediately moves towards C.

a Use the symbolism, F(by ... on ...), to identify the force by C and the force by D.

_____

_____

_____

b Describe, in terms of force, the effect that C has on D and D has on C.

_____

_____

_____

c Is it possible to decide the charge on each balloon? Explain.

_____

_____

_____

5 A child sits quietly on a trampoline.

a Use the symbolic force nomenclature, $\vec{F}$(by… on …), to describe the force applied:

i on the child

_____

_____

_____

ii on the trampoline

_____

_____

_____

iii by the weight force applied on the child.

_____

_____

_____

b Use symbols from Question **5a** to write a vector equation to describe the sum of forces on the child.

_____

_____

_____

9780170412551

6 **a** Explain the difference between mass and weight.

_____

_____

_____

b Explain how the weight of a mass on Earth and the weight of the same mass on the Moon could be made equal.

_____

_____

_____

7 A 1.5t container of essential equipment is sent from Earth to the Mars Colony. The spring scale used to measure this mass is also sent to Mars. The colonists wonder why they only received 566 kg of equipment. Explain to the colonists, in quantitative detail, what happened, if anything, to rest of the equipment.

_____

_____

_____

_____

_____

_____

_____

_____

_____

_____

_____

_____

_____

_____

_____

_____

_____

_____

14.2 | Newton's three laws

WORKED EXAMPLES

1 An object of mass 6.0 kg is pushed forwards by a constant force of 25 N. It travels at constant speed of $5.0\,\mathrm{m\,s^{-1}}$. Explain how this can happen.

ANSWER

Constant speed means that the acceleration is zero.

From Newton's second law:

If $\vec{a} = 0$, then $\sum \vec{F} = 0$

$F(\text{forwards}) - F(\text{backwards}) = 0\,\mathrm{N}$

$\rightarrow$ $25\,\mathrm{N} - F(\text{backwards}) = 0\,\mathrm{N}$

$\rightarrow$ $F(\text{backwards}) = 25\,\mathrm{N}$

$\rightarrow$ The body continues in its state of uniform motion because the backwards directed force is 25 N, so that the net force on the body is zero.

2 Find the acceleration of a 500 kg mass when it is subjected to a net force of 50 N.

ANSWER

From Newton's second law:

$$\vec{a} = \frac{\sum \vec{F}}{m}$$

$$\Rightarrow \vec{a} = \frac{50\,\mathrm{N}}{500\ \mathrm{kg}}$$

$$\Rightarrow a = 0.1\,\mathrm{m\,s^{-2}}$$

3 The graph (Figure 14.2.1) shows the variation of speed with time of a 0.75 kg model train. The train applies a maximum forwards driving force, D, of 4.0 N. The model car is subject to a constant resistance force, R.

FIGURE 14.2.1

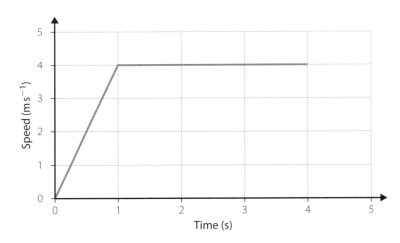

a During the first 1.0 s, calculate:

   i the acceleration of the vehicle

   ii the net force on the vehicle

   iii the resistance force, R.

b Calculate the driving force at $t = 3.0\,s$.

9780170412551

ANSWER

a **i** $a = \dfrac{v}{t}$

$\Rightarrow a = \dfrac{4.0 \text{ m s}^{-1} - 0 \text{ m s}^{-1}}{1.0 \text{ s} - 0.0 \text{ s}}$

$\Rightarrow a = 4.0 \text{ m s}^{-2}$

ii $\sum F = ma$

$\Rightarrow \sum F = 0.75 \text{kg} \times 4.0 \text{ m s}^{-2}$

$\Rightarrow \sum F = 3.0 \text{ N}$

iii $\sum F = D - R$

$\Rightarrow R = D - \sum F$

$\Rightarrow R = 4.0 \text{ N} - 3.0 \text{ N}$

$\Rightarrow R = 1.0 \text{ N}$

b At $t = 3.0 s$ the net force on the train is zero (constant speed; Newton's first law). The driving force on the train is equal to the resistance force, that is, 1.0N.

4 A horse pushes into the harness in order to apply a force on the poles attached to the cart. The cart pulls along the poles with exactly the same force in the opposite direction. How then can the horse get the cart to move? Explain.

ANSWER

It is the net force on the cart that causes the cart to move forward. The horse must be able to apply a net force on the cart in the forward direction. This forward force is applied by the horse on the poles as well as by the ground on the horse's feet, which is part of the reaction force to the horse's attempt to push Earth backwards.

QUESTIONS

1 The graph (Figure 14.2.2) shows the variation of speed with time of a 2.0t vehicle. The vehicle applies a forwards driving force, D. The vehicle is subject to a combined resistant force, R, of 2000N.

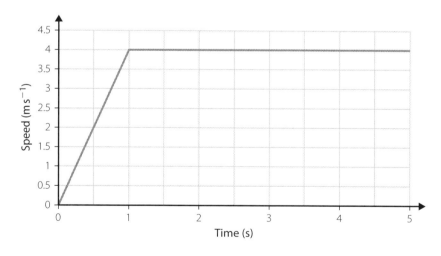

FIGURE 14.2.2

a At time, $t = 5.0$ s, calculate:

i the acceleration of the vehicle

_____

_____

ii the net force on the vehicle

_____

_____

iii the forwards driving force, D.

_____

_____

b Explain what happens to the driving force at $t = 10$ s.

_____

_____

_____

2 Complete the following table.

MASS (kg)	NET FORCE ON MASS (N)	ACCELERATION (m s^{-2})
2.5	15	
1.68×10^4	3.012	
0.127		6.3743
	852	9.138

3 The normal force and weight are not an action–reaction pair. Explain.

_____

_____

_____

_____

4 In a tug-of-war, the force on the rope applied by team A, pulling to the left, is equal to the force on the rope applied by team B pulling to the right. These two forces are equal in size and opposite in direction. Explain how one team can pull the other team over the line.

_____

_____

_____

_____

5 An athlete performs a standing jump by pushing down on the ground. According to Newton's third law, the ground pushes up with the same magnitude of force. Since these two forces are equal in size and opposite in direction, how is it possible to jump up?

_____

_____

_____

_____

9780170412551

14.3 | Free-body diagrams

WORKED EXAMPLE

Two people start to push a 30 kg box along the floor. One person applies a 10 N force to the right, the other a 20 N force also to the right. There is a constant 25 N friction force. Find the time it takes for the box to be travelling at $1.5\,\mathrm{m\,s^{-1}}$.

ANSWER

1 Free-body diagram including direction of the net force and acceleration:

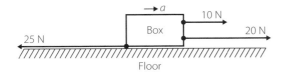

2 Newton's second law applies:

$$a = \frac{\sum F}{m}$$

$$\Rightarrow a = \frac{10\ \mathrm{N} + 20\ \mathrm{N} - 25\ \mathrm{N}}{30\ \mathrm{kg}}$$

$$\Rightarrow a = \frac{5\ \mathrm{N}}{30\ \mathrm{kg}}$$

$$\Rightarrow a = 0.167\ \mathrm{m\,s^{-2}}$$

3 Solve using kinematic formula:

$$v = u + at$$

$$\Rightarrow t = \frac{v - u}{a}$$

$$\Rightarrow t = \frac{1.5\ \mathrm{m\,s^{-1}} - 0.0\ \mathrm{m\,s^{-1}}}{0.167\ \mathrm{m\,s^{-2}}}$$

$$\Rightarrow t = 9.0\ \mathrm{s}$$

QUESTIONS

1 An ice-skater (mass = 50 kg) is accelerated at $1.5\,\mathrm{m\,s^{-1}}$ by a combination of the forward force applied by the ice to the skates (40 N) and a friend (45 N). Calculate the friction force applied by the ice.

_____

_____

_____

2 Calculate the average force applied by a trampoline to launch a 65 kg person at a vertical speed of $12\,\mathrm{m\,s^{-1}}$ in 1.2 s.

_____

_____

_____

3 A crane cable accelerates a 2.7-tonne load from rest to $1.8\,\mathrm{m\,s^{-1}}$ over a distance of 3.4 m. Later, the load travels at a constant speed of $2.15\,\mathrm{m\,s^{-1}}$. Calculate the force applied by the cable when the load is travelling at:

a $0.74\,\mathrm{m\,s^{-1}}$

_____

_____

_____

b $2.15\,\mathrm{m\,s^{-1}}$.

_____

_____

_____

4 A 55 kg gymnast performs a vertical jump by pushing on the floor for 155 ms. She leaves the floor at $5.5\,\mathrm{m\,s^{-1}}$.

a Calculate the average acceleration of the gymnast.

_____

_____

_____

b Find the average net force applied by the floor to the gymnast.

_____

_____

_____

c Determine the average force applied by the gymnast to the floor.

_____

_____

_____

5 Three blocks, P, Q and R, of mass 4.0 kg, 5.0 kg and 6.0 kg respectively are touching. An external force of 15 N is applied to P.

a Find the acceleration of P.

_____

_____

b **i** Find $\vec{F}$ (by Q on R)

_____

_____

ii Find $\vec{F}$(by Q on P)

_____

_____

iii Find $\Sigma \vec{F}$(on Q)

_____

_____

14.4 Conservation of momentum

Momentum change is caused when a force acts over a time interval. The action of a force over time is called the impulse, $\vec{J}$: Impulse causes momentum change. For the law of conversation of momentum, the total momentum in an isolated system is always the same.

▶ The total momentum before, during and after a collision is the same:

$$\Sigma \vec{p}_{\text{before}} = \Sigma \vec{p}_{\text{after}}$$
$$\Rightarrow \Sigma (m\vec{v})_{\text{before}} = \Sigma (m\vec{v})_{\text{after}}$$

The conservation of momentum law is a direct consequence of Newton's third law:

▶ Throughout an interaction between two bodies, the Newton 3 pair of forces act simultaneously over the same time interval to change the momentum of the interacting objects

▶ The impulse by the first object on a second object is the same as the impulse by the second object on the first object.

WORKED EXAMPLES

1 A 3.0 kg mass is subject to a force of 4.0 N for 5.0 s. Calculate:

   a the impulse of the force on the mass

   b the change in momentum of the mass

   c the change in speed of the mass

ANSWER

a $J = F \Delta t$

   $\Rightarrow J = 4.0 \text{ N} \times 5.0 \text{ s}$

   $J = 20 \text{ N s}$

b $J = \Delta p = 12 \text{ N s}$

c $\Delta p = m\Delta v$

   $\Rightarrow \Delta v = \dfrac{\Delta p}{m}$

   $\Rightarrow \Delta v = \dfrac{12 \text{ N s}}{3.0 \text{ kg}}$

   $\Rightarrow \Delta v = 4.0 \text{ m s}^{-1}$

2 A tram of mass 23.0 tonne travelling at $24.0\,\text{m s}^{-1}$ collides with a stationary tram of mass 22.0 tonne. The two trams move off together. Find the speed of the 23.0 tonne tram after the collision.

ANSWER

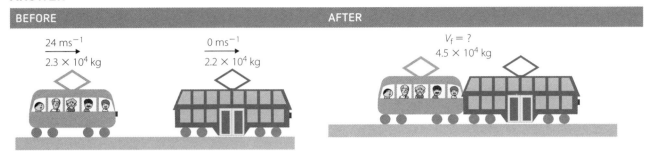

Equation:

$$\vec{P}_\text{T}(\text{before}) = \vec{p}_1 + \vec{p}_2$$
$$\Rightarrow \vec{P}_\text{T}(\text{before}) = m_1 v_1 + m_2 v_2$$
$$\Rightarrow \vec{P}_\text{T}(\text{before}) = 2.30 \times 10^4\,\text{kg} \times {}^+24.0\,\text{m s}^{-1}$$
$$+\, 2.20 \times 10^4\,\text{kg} \times 0.0\,\text{m s}^{-1}$$

$$\Rightarrow \vec{P}_\text{T}(\text{before}) = {}^+5.52 \times 10^5\,\text{kg m s}^{-1}$$

Equation:

$$\vec{P}_\text{T}(\text{after}) = \vec{p}_{(1+2)}$$
$$\Rightarrow \vec{P}_\text{T}(\text{after}) = (m_1 + m_2) v_f$$
$$\Rightarrow \vec{P}_\text{T}(\text{after}) = (2.30 \times 10^4\,\text{kg} + 2.20 \times 10^4\,\text{kg}) \times v_f$$

$$\Rightarrow \vec{P}_\text{T}(\text{after}) = {}^+4.50 \times 10^4\,\text{kg} \times v_f$$

$$\vec{P}_\text{T}(\text{before}) = \vec{P}_\text{T}(\text{after})$$
$$\Rightarrow {}^+5.52 \times 10^5\,\text{kg m s}^{-1} = {}^+5.50 \times 10^4\,\text{kg} \times v_f$$
$$\Rightarrow v_f = \frac{{}^+5.52 \times 10^5\,\text{kg m s}^{-1}}{{}^+4.50 \times 10^4\,\text{kg}}$$
$$\Rightarrow v_f = 1.23\,\text{m s}^{-1}$$

The speed of the 23.0 tonne tram after the collision is the same as the speed of the combined trams, $1.23\,\text{m s}^{-1}$.

QUESTIONS

1 Complete the following table:

MASS (kg)	SPEED (m s⁻¹)	MOMENTUM (kg m s⁻¹)
4.0	6.0	
	13.7	127.8
3.43×10^5		2.18×10^6

2 Complete the following table:

MASS (kg)	TIME INTERVAL (s)	FORCE (N)	MOMENTUM CHANGE (kg m s⁻¹)	SPEED CHANGE (m s⁻¹)
8.0	4.0	6.0		
	2.6	19.6		8.67
		9.77×10^5	4.61×10^6	26.5
2.4×10^3	10.0		13.7×10^4	

9780170412551

3 A 212 N force is applied for 18.6 s on a 13.7 kg mass that is travelling at $4.15\,\text{m}\,\text{s}^{-1}$.

a Calculate the impulse of the force on the mass.

_____

_____

_____

b Calculate the change in momentum of the mass.

_____

_____

_____

c Determine the final speed of the mass.

_____

_____

_____

4 A billiard ball travelling at $2.0\,\text{m}\,\text{s}^{-1}$ collides with a stationary billiard ball. Prove that all the momentum of the moving ball is transferred to the stationary ball.

_____

_____

_____

_____

5 A train carriage of mass 345 kg travelling at $4.71\,\text{m}\,\text{s}^{-1}$ collides with a similar carriage travelling in the opposite direction at $5.65\,\text{m}\,\text{s}^{-1}$. They stick together. Find the speed of the combined carriages after the collision.

_____

_____

_____

_____

6 A proton travelling at $3.60\times10^{4}\,\text{m}\,\text{s}^{-1}$ collides with a stationary helium nucleus. Find the speed of the proton and the helium nucleus after the collision. (A helium nucleus is, to a good approximation, four times more massive than a proton).

_____

_____

_____

_____

EVALUATION

1 Which pair of forces includes two different types of non-contact forces?

   A Gravity and mass

   B Gravitational force and weight

   C Magnetic force and normal force

   D Electrostatic force and normal force

2 Which of the following statements is always true for a moving object?

   A A net force on the object causes it to move at constant speed.

   B If the object experiences external forces its speed must change.

   C A net external force will always cause the object's velocity to change.

   D The object will eventually stop, even if no external forces are applied.

3 If the net force on a moving mass is constant, but not zero:

   A its speed will be constant.

   B its velocity will be constant.

   C its acceleration will be constant.

   D its momentum will be constant.

4 What is the correct way to state the weight of a 10 kg mass near Earth's surface?

   A 10 kg

   B 98 kg

   C 10 N

   D 98 N

5 In which place would you have the least weight?

   A 100 km above the surface of Earth

   B Standing on the surface of the Moon

   C 100 km above the surface of the Moon

   D None of these (weight is constant wherever you are)

9780170412551

6 The driver of a car is wearing a seatbelt when the car crashes into a tree. The driver comes to a stop inside the car. What can be said about the driver's acceleration during this time?

A The driver continues to accelerate in the direction of motion.

B The driver accelerates in the opposite direction to the motion.

C The acceleration of the driver is positive.

D The deceleration of the driver is zero.

7 The acceleration of a point mass can be maximised by:

A increasing the net force and decreasing the mass.

B increasing the mass and decreasing the net force.

C increasing both the net force and the mass.

D decreasing both the net force and the mass.

8 A horizontal force of 250 N is applied to a 20 kg mass sitting on a level surface. A frictional force of 90 N acts on the mass. The magnitude of the acceleration is:

A 8.0m s^{-2}

B 12.5m s^{-2}

C 16.0m s^{-2}

D 17.0m s^{-2}

9 A book sits on a horizontal table without moving. Select the pair of forces that represent Newton's third law.

A Gravitational force on book by Earth's mass; normal force on book by table

B Gravitational force by book on Earth's mass; gravitational force by Earth's mass on book

C Normal force by table on book; gravitational force by book on table

D Normal force by book on Earth's mass; gravitational force by Earth's mass on book

10 What is the momentum of a 3.6 kg mass travelling at 2.5m s^{-1} is closest to?

A 9.0kg m s^{-2}

B 10kg m s^{-1}

C 9.0N s^{-1}

D 10N m s^{-1}

11 A 4.5 kg mass at rest is accelerated by a force that acts for 5.0 s. It reaches a speed of 6.0m s^{-1}. What is the magnitude of the impulse of the force?

A 22.5N s

B 27.0N s

C 30.0kg m s^{-1}

D There is insufficient data to calculate the impulse.

12 A 7.0kg mass travelling at $3.0\,\mathrm{ms}^{-1}$ is subject to a force of 14.0N for 21.0s. What is the change of momentum?

A $21\,\mathrm{kgms}^{-1}$

B $294\,\mathrm{kgms}^{-1}$

C $441\,\mathrm{kgms}^{-1}$

D $2.1\times10^{3}\,\mathrm{kgms}^{-1}$

13 A 2.0kg box slides over a frictionless surface at $^{+}3.0\,\mathrm{ms}^{-1}$. It collides with a 5.0kg stationary box, which moves off at $^{+}2.0\,\mathrm{ms}^{-1}$. After the collision, what is the velocity of the 2.0kg box?

A $^{-}4.0\,\mathrm{ms}^{-1}$

B $^{+}4.0\,\mathrm{ms}^{-1}$

C $^{-}2.0\,\mathrm{ms}^{-1}$

D $^{+}2.0\,\mathrm{ms}^{-1}$

14 A force of 2.0×10^{5} N is applied to a baseball of mass 140g for 0.80ms. The baseball leaves the bat at $60\,\mathrm{ms}^{-1}$. What impulse is applied to the baseball by the bat?

A $8.4\,\mathrm{kgms}^{-1}$

B $8.4\times10^{3}\,\mathrm{kgms}^{-1}$

C $160\,\mathrm{Ns}$

D $1.6\times10^{5}\,\mathrm{Ns}$

15 A 2.37×10^{4} N force is applied for 2.56ms on a 13.7kg mass, which is travelling at $4.15\,\mathrm{ms}^{-1}$. Calculate the final speed of the mass.

_____

_____

_____

_____

_____

16 A person pushes a 650kg boat horizontally with a 200N force. The boat is also pulled by a winch, which applies a 850N force. There is a constant 350N friction force. Find the time it takes for the boat to be travelling at $1.2\,\mathrm{ms}^{-1}$.

_____

_____

_____

_____

_____

17 The graph (Figure 14.5.1) shows the variation of speed with time of a 450 kg motorcycle. The motorcycle applies a forwards driving force, D. The combined resistance forces total 1500 N.

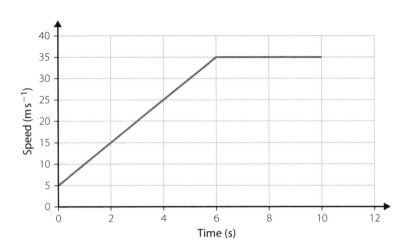

FIGURE 14.5.1

At time, $t = 2.0$ s, calculate:

a the acceleration of the vehicle

_____

_____

_____

b the forward driving force, D

_____

_____

_____

c the driving force at $t = 8.0$ s.

_____

_____

_____

18 A train carriage of mass 1.50×10^3 kg travelling at $9.0\,\text{m s}^{-1}$ collides with a stationary carriage of mass 1.00×10^3 kg. The two carriages move off together. Find the speed of the 1.00 tonne carriage after the collision.

_____

_____

_____

_____

_____

15 Newton's laws of motion

LEARNING

Summary

▸ When forces act to move an object over a distance, work is done. Work can be done on an object to move it by some distance. Alternatively, work can be done by an object to move some mass over a distance. Thus, work transfers energy either to or from a system.

▸ An isolated system is a system that has no energy transfers into or out of it, and no energy is created or destroyed inside the system.

▸ The total energy in an isolated system is constant. There is no change to the total energy.

▸ A non-isolated system can have energy added or removed by work done on or by the system. If the energy changes from an initial value E_i to final value E_f, the difference in energy or change in energy is:

$$\Delta E = E_f - E_i$$

where E_i = initial energy state

E_f = final energy state

▸ In an isolated system, the law of conservation of energy can be re-stated as:

$$\Delta E = 0$$

$$\rightarrow E_f - E_i = 0$$

$$\rightarrow E_f = E_i$$

▸ Work transfers energy:

$$W = F_{\parallel}s$$

Unit: newton metre (N.m) $= \text{kg m}^2\,\text{s}^{-2} =$ joule, J

▸ The force and the distance interval must be in the same direction:

$$W = F_{\parallel}s$$

where W = work done

$F_{\parallel}$ = force parallel to the distance

▸ Energy comes in two basic forms: kinetic energy and potential energy.

▸ Kinetic energy: energy of motion:

$$E_K = \frac{1}{2}mv^2$$

9780170412551

- Change of kinetic energy is given by:

$$\Delta E_K = \frac{1}{2}mv_f^2 - \frac{1}{2}mv_i^2$$

$$\Rightarrow \Delta E_K = \frac{1}{2}m(v_f^2 - v_i^2)$$

- Potential energy is energy stored in a system, ready to be returned to the system in another form.
- Work done by a constant force in the direction of the displacement:

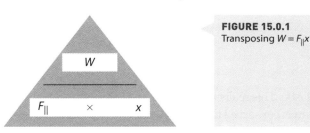

FIGURE 15.0.1
Transposing $W = F_\| x$

- Three types of questions can be identified:
  - Find W given $F_\|$ and x: $W = F_\| x$

  - Find $F_\|$ given W and x: $F_\| = \dfrac{W}{x}$

  - Find x given $F_\|$ and W: $x = \dfrac{W}{F_\|}$

- Work done by a stepwise force in the direction of displacement:

$$W = F_\| x$$

- Calculate the sum of the work done by each force.
- Work done by a component of force in the direction of displacement:

$$W = F_\| s = Fs\cos\theta$$

- The amount of work done on an object can result in:
  - kinetic energy increasing or decreasing:

$$W = \Delta E_k = \Delta\left(\frac{1}{2}mv^2\right)$$

AND/OR
  - potential energy being stored in a system ready to do work:

$$W = \Delta E_p$$

- Energy is stored in systems such as springs and Earth's gravitational field.
  - A spring can be considered to be a system. When work is done on a spring to stretch or compress it, potential energy is stored in the spring as a system.
  - When the spring system does work on a mass, the spring system converts the stored potential energy to kinetic energy. A spring acts to oppose any change to its length. The force by a spring is, therefore, a restoring force because it acts to restore the spring to its original length. Hooke's law concerns the property of a spring:

$$F(\text{by spring}) \propto -x$$

$$\Rightarrow F = k(-x)$$

$$\Rightarrow F = -kx$$

  - Elastic potential energy is the energy stored in a spring system.
  - A spring provides a constantly increasing, restoring force of magnitude kx, where k represents the stiffness of the spring. The work done on the spring system is the energy stored in the spring system:

$$E_p = \frac{1}{2}kx^2$$

- The gravitational field can be considered to be a system. When work is done on a mass by working against the gravitational field, potential energy is stored in the field as a system.
  - Near Earth, Earth's gravitational field is a constant, g.
  - The work done on the system is the energy stored by a constant force, mg, over a distance, h.
    $E_p = mgh$
  - When the gravitational field does work on a mass, the field converts the stored potential energy to kinetic energy. If the potential energy $E_p = 0$, at height $h = 0$:

    $$v_{max} = \sqrt{2gh_{max}} \quad \text{or} \quad h_{max} = \frac{v_{max}^2}{2g}$$

    $$mgh = \frac{1}{2}mv^2$$

- Collisions can be elastic or inelastic. This refers to the kinetic energy before and after the collision.
  - In every collision momentum is conserved throughout, but not so kinetic energy.
  - Elastic and inelastic refer to the total kinetic energy in a system:
    - Elastic collisions: $E_k(\text{before}) = E_k(\text{after})$
      Elastic collisions return any potential energy stored during the collision.
    - Inelastic collisions: $E_k(\text{before}) \neq E_k(\text{after})$
      Inelastic collisions retain some potential energy within the system (e.g. re-arrangements of molecular bond energies) and send some energy out of the system (e.g. heat, sound, light).
  - In all collisions in closed systems, momentum is conserved throughout the collision. This is a consequence of Newton's third law, which states that forces of interaction, *which must be of the same type*, always:
    - are equal in magnitude
    - are opposite in direction
    - act on different objects.
      $$\Delta p_t = 0$$

9780170412551

15.1 Law of conservation of energy

WORKED EXAMPLES

1 When a constant force of 5.0 N is exerted to push a 2.0 kg mass a distance of 1.5 m, find:

   a the work done

   b the amount of energy transferred.

ANSWER

a $W = F_{\parallel}x$

    $\Rightarrow W = 5.0 \text{ N} \times 1.5 \text{ m}$

    $\Rightarrow W = 7.5 \text{ J}$

b $\Delta E = 7.5 \text{ J}$

2 Figure 15.1.1 shows how the net force varies as a 200 kg object is being pushed along a horizontal floor.

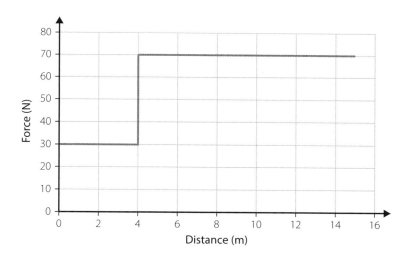

FIGURE 15.1.1

   a Calculate the energy transferred in the first 4.0 m.

   b Find the work done when the object is moved 10 m.

ANSWER

a $W = F_{\parallel}x$

$\Rightarrow W = 30\,\text{N} \times (4.0\,\text{m} - 0.0\,\text{N})$

$\Rightarrow W = 120\,\text{J}$

$\Rightarrow \Delta E = 120\,\text{J}$

b $W = W_1 + W_2$

$\Rightarrow W = F_1 x_1 + F_2 x_2$

$\Rightarrow W = 30\,\text{N} \times (4.0\,\text{m} - 0.0\,\text{m}) + 70\,\text{N} \times (10.0\,\text{m} - 4.0\,\text{m})$

$\Rightarrow W = 120\,\text{J} + 420\,\text{J}$

$\Rightarrow W = 540\,\text{J}$

3 A net force of 10 N acting at an angle of 60° to the horizontal pushes a 7.0 kg box a horizontal distance of 4.0 m.

    a Calculate the work done.

    b Find the speed of the box after 4.0 m.

ANSWER

a $W = F_{\parallel}x$

$\Rightarrow W = 10\,\cos 60°\,\text{N} \times 4.0\,\text{m}$

$\Rightarrow W = 20\,\text{J}$

b $\Delta E_k = \dfrac{1}{2}mv^2 = W$

$\Rightarrow v = \sqrt{\dfrac{2W}{m}}$

$\Rightarrow v = \sqrt{\dfrac{2 \times 20\,\text{J}}{7.0\,\text{kg}}}$

$\Rightarrow v = 5.7\,\text{m s}^{-1}$

9780170412551

QUESTIONS

1 Complete the table.

$F_{\parallel}$ (N)	x (m)	W (J)
20	10	
	0.76	18.04
2.21×10^5		6.87×10^3

2 Find the distance through which a car is moved by a force of 5.0 N when 25 kJ of work is done on the car.

_____

_____

_____

3 Figure 15.1.2 shows how the net force varies while a 6.0 t container is being pushed along a horizontal floor by a forklift. There is a constant frictional force of 4000 N.

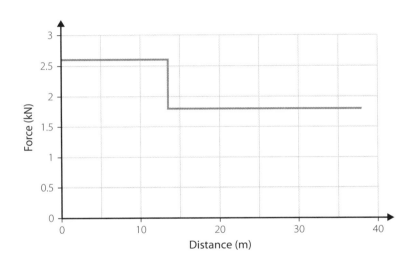

FIGURE 15.1.2

a Find the work done by the forklift in pushing the load a distance of:

i 5.0 m

_____

_____

ii 32.5 m.

_____

_____

b Calculate the speed of the container after it has moved 30 m.

4 Complete the following table.

FORCE (N)	ANGLE RELATIVE TO DISPLACEMENT (°)	FORCE PARALLEL TO DISPLACEMENT (N)	DISTANCE (m)	WORK (J)
40	30		10	
0.359	10		2.3×10^{-3}	
9.72×10^3	42.7		17.89	
	60		5.00	35
8.64			4.17	7.31

5 A shopper and child push a 15 kg trolley for 5.0 m along a supermarket lane. They both hold on to the same position on the front bar as they push. The shopper pushes down on the bar with a force of 12 N at an angle of 45° to the horizontal. The child pushes up with a force of 5.0 N at an angle of 60° to the horizontal. The combined, horizontal resistance to the motion is 4.0 N.

a Draw the free-body diagram.

b On the free-body diagram, show the horizontal and vertical components of the forces applied to the trolley.

c Produce a scale drawing to show the vector sum of all forces on the trolley.

9780170412551

d Calculate the work done on the trolley.

_____

_____

_____

_____

15.2 | Energy stored in systems

WORKED EXAMPLES

1 A spring stretches by 15 cm when a force of 90 N is applied.

    a Find the spring constant.

    b Calculate the energy stored when the spring is stretched by 8.5 cm.

ANSWER

a $F = kx$

$$\Rightarrow k = \frac{F}{x}$$

$$\Rightarrow k = \frac{90 \text{ N}}{0.15 \text{ m}}$$

$$\Rightarrow k = 6.0 \times 10^2 \text{ N m}^{-1}$$

b $E_P = \frac{1}{2}kx^2$

$$\Rightarrow E_P = \frac{1}{2} \times 6.0 \times 10^2 \text{ N m}^{-1} \times (0.085 \text{ m})^2$$

$$\Rightarrow E_P = 2.17 \text{ J}$$

2 A crane lifts a 5.0 t mass from the ground to a vertical height of 30 m. The mass then falls from this height back to ground.

    a Calculate the potential energy stored in the field at 30 m.

    b Calculate the speed at the moment the mass passes a point 10 m above the ground.

ANSWER

a $E_p = mgh$

$$\Rightarrow E_p = 5.0 \times 10^3 \text{ kg} \times (30 \text{ m} - 0 \text{ m})$$

$$\Rightarrow E_p = 1.5 \times 10^5 \text{ J}$$

b $v_{max} = \sqrt{2gh}$

$$\Rightarrow v_{max} = \sqrt{2 \times 9.80 \text{ ms}^{-2} \times 30 \text{ m}}$$

$$\Rightarrow v_{max} = 24.2 \text{ ms}^{-1}$$

QUESTIONS

1 Compare the stiffness of spring S, which applies a force of 8.36 N when it is extended by 4.12 cm, and the stiffness of spring T, which extends by 5.03 m when a force of 7.32×10^2 N is applied to one end.

_____

_____

_____

2 Complete the following table:

FORCE APPLIED TO SPRING (N)	FORCE APPLIED BY SPRING (N)	EXTENSION (m)	STIFFNESS (N m^{-1})	HOOKE'S LAW EQUATION	POTENTIAL ENERGY GAIN (J)
548		0.341			
	45			$F=-9.14x$	
		1.27			67.9
	1947				

3 The graph (Figure 15.2.1) shows how the force applied by an open-coiled spring varies with the extension of the spring.

FIGURE 15.2.1

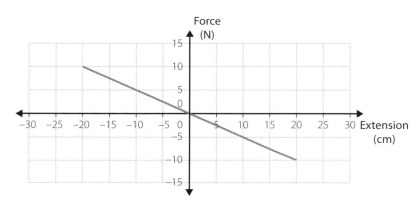

Use the graph to find, in SI units:

a the spring constant or stiffness

_____

_____

_____

b the energy stored in the spring when it is:

i extended by 10 cm

_____

_____

_____

ii compressed by 15 cm.

_____

_____

_____

c A 25 g ball is placed next to the spring when it is compressed by 20 cm. When the spring is released, the ball is launched. Find the speed of the ball at release.

_____

_____

_____

4 A spring was extended by an external force. The following data were obtained.

EXTENSION FORCE APPLIED (N)	LENGTH OF SPRING (cm)	SPRING EXTENSION (cm)
0.0	15.5	0.0
2.5	21.8	
4.8	27.2	
7.6	44.1	
8.3	36.0	
9.8	40.5	

a Plot the following graphs, including the line of best fit for the data:

i extension as a function of the external force applied

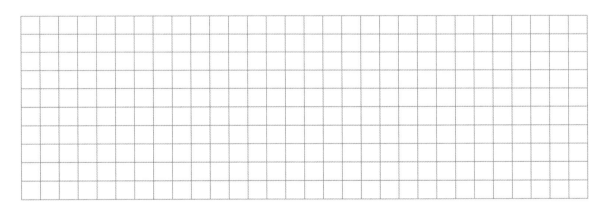

ii force applied by spring as a function of extension.

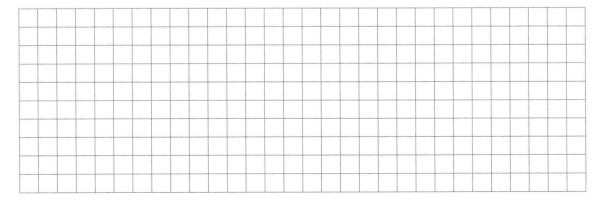

b **i** State Hooke's law.

_____

_____

ii Define all terms used.

_____

_____

_____

iii Which graph, **a i** or **a ii**, expresses Hooke's law correctly? Explain in terms of one of Newton's laws.

_____

_____

_____

c **i** Find the extension that causes the spring to exert a force of 6.0 N.

_____

_____

_____

ii Find the stiffness of the spring in SI units.

_____

_____

_____

iii Find the energy stored in the spring at maximum extension.

_____

_____

_____

d Using the data provided, is it appropriate to predict the force exerted by the spring for an extension of 80 cm? Explain your reasoning.

_____

_____

_____

_____

5 Complete the table below, which shows data for different masses that are lifted vertically above the ground and allowed to drop. ($g = 9.80\,\mathrm{m\,s^{-1}}$)

MASS (kg)	WEIGHT (N)	VERTICAL DISTANCE (m)	MAXIMUM POTENTIAL ENERGY (J)	MAXIMUM KINETIC ENERGY (J)	MAXIMUM SPEED (m s⁻¹)
254		45			
	1050		3.67×10^3		
31.5					7.9
		6.83	9.87×10^5		

WORKED EXAMPLE

The graph (Figure 15.2.2) shows how the force of gravity on a 1.0 kg mass near the surface of Mars varies with distance from the surface.

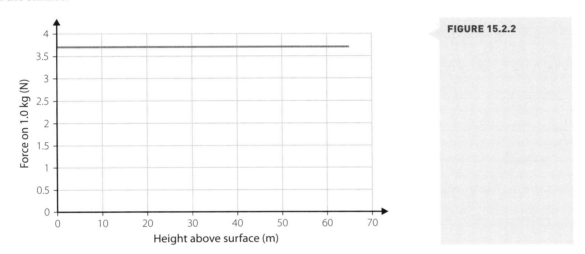

FIGURE 15.2.2

Calculate the potential energy gained when a 10 kg mass is moved from 15 m to 35 m above Mars.

ANSWER

$\Delta E_\mathrm{P} = F(\text{on } 10\,\mathrm{kg}) \times \Delta h$

but $F(\text{on } 10\,\mathrm{kg}) = 10\,\mathrm{kg} \times 3.7\,\mathrm{N\,kg^{-1}}$

$\Rightarrow F(\text{on } 10\,\mathrm{kg}) = 37\,\mathrm{N}$

$\Rightarrow \Delta E_\mathrm{P} = 37\,\mathrm{N} \times (35\,\mathrm{m} - 15\,\mathrm{m})$

$\Rightarrow \Delta E_\mathrm{P} = 740\,\mathrm{N}$

QUESTIONS

1   The graph (Figure 15.2.3) shows the force applied to a 1.0 kg mass near the surface of Jupiter. A 30 kg mass is lifted vertically from 5.0 m near the surface of Jupiter. When it is dropped, it then returns to its original position.

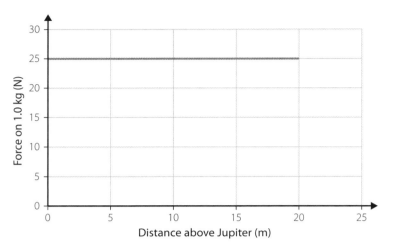

FIGURE 15.2.3

a What is the weight of the mass?

_____

_____

_____

b Calculate the energy needed to lift the mass from 5.0 m to 20 m.

_____

_____

_____

c Calculate the speed of the mass at 5.0 m when it is dropped from 25 m.

_____

_____

_____

2   Use Earth's gravitational acceleration and the graphs above, which show how the force of gravity on a 1.0 kg mass near the surfaces of Mars and Jupiter respectively vary with distance from the surface, to answer the following questions.

a Calculate the energy needed to launch a 2.5 kg projectile to a height of 45 m above:

i Mars

_____

_____

_____

9780170412551

ii Earth

_____

_____

_____

iii Jupiter.

_____

_____

_____

b Calculate the minimum speed necessary to launch a 45.3 kg projectile from a height of 9.0 m to a height of 47 m above:

i Mars

_____

_____

_____

ii Earth

_____

_____

_____

iii Jupiter.

_____

_____

_____

15.3 | Elastic and inelastic collisions

Analysis of collisions is enhanced by setting out all the details in *Before* and *After* columns:

BEFORE	AFTER
Sketch to show situation before collision Show all data, incl. direction Name all unknown variables Equation for total momentum before: p_T(before)	Sketch to show situation after collision Show all data, incl. direction Name all unknown variables Equation for total momentum after: p_T(after)
p_T(before) = p_T(after)	
Equation for total kinetic energy before: E_k(*total before*)	Equation for total kinetic energy before: E_k(*total after*)

Compare:

$$E_k\text{(total before) and } E_k\text{(total after)}$$

WORKED EXAMPLE

A 4.0 kg mass travelling to the right at $3.5\,\mathrm{m\,s^{-1}}$ collides with a 2.0 kg mass moving to the left at $4.0\,\mathrm{m\,s^{-1}}$. The masses stick together and move off to the right at a speed, v_f.

a Find the final velocity of the combined mass.

b Use calculations to show that the collision is inelastic.

ANSWER

BEFORE

AFTER

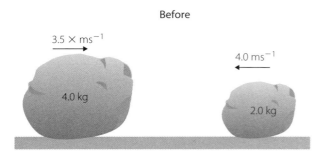

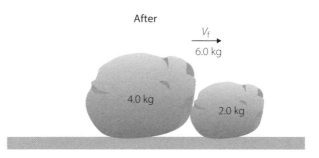

$p_T(\text{before}) = {}^+(4.0\text{ kg} \times 3.5\text{ m s}^{-1}) + ({}^-2.0\text{ kg} \times 4.0\text{ m s}^{-1})$

$\Rightarrow p_T(\text{before}) = {}^+6.0\text{ kg m s}^{-1}$

$p_T(\text{after}) = {}^+(4.0\text{ kg} + 2.0\text{ kg}) \times v_f$

$\Rightarrow p_T(\text{after}) = {}^+6.0\text{ kg} \times v_f$

$p_T(\text{before}) = p_T(\text{after})$

${}^+6.0\text{ kg} \times v_f = {}^+6.0\text{ kg m s}^{-1}$

$\Rightarrow v_f = {}^+1.0\text{ m s}^{-1}$

$E_k(\text{total before}) = \dfrac{1}{2} \times 4.0\text{ kg} \times (3.5\text{ m s}^{-1})^2$

$+ \dfrac{1}{2} \times 2.0\text{ kg} \times (4.0\text{ m s}^{-1})^2$

$\Rightarrow E_k(\text{total before}) = 40.5\text{ J}$

$E_k(\text{total after}) = \dfrac{1}{2} \times (4.0\text{ kg} + 2.0\text{ kg}) \times (1.0\text{ m s}^{-1})^2$

$\Rightarrow E_k(\text{total after}) = 3.0\text{ J}$

Compare:
$E_k\text{(total before)} \neq E_k\text{(total after)}$
$\Rightarrow$ inelastic

9780170412551

QUESTIONS

1 A 14 kg mass travelling to the right at $1.5\,\text{m s}^{-1}$ collides with a 21.0 kg mass moving to the left at $3.0\,\text{m s}^{-1}$. They stick together and move off at speed, v_f.

   a Find the final velocity, v_f, of the combined mass.

   b Use calculations to show that the collision is inelastic.

2 A train carriage of mass 2.3 t runs up to another stationary carriage of mass 1.8 t at a speed of $6.0\,\text{m s}^{-1}$. The two carriages couple and move off together.

   a Find the final velocity, v_f, of the combined mass.

   b Use calculations to show that the collision is inelastic.

3 A fast neutron of mass $1.67 \times 10^{-27}\,\text{kg}$ travels at $2.2 \times 10^{7}\,\text{m s}^{-1}$ in a nuclear pile. It collides with a stationary deuterium atom of mass $3.34 \times 10^{-27}\,\text{kg}$ and bounces off at $2.2 \times 10^{3}\,\text{m s}^{-1}$.

   a Find the final velocity, v_f, of the deuterium atom.

   b Use calculations to decide if the collision is elastic or inelastic.

4 A 4.4×10^3 kg coal truck travels at $6.3\,\mathrm{m\,s^{-1}}$ under a chute that drops $4.6\,\mathrm{t}$ of coal vertically into the truck.

a Find the final velocity, v_f, of the coal truck.

_____

_____

_____

b Use calculations to decide if the collision is elastic or inelastic.

_____

_____

_____

9780170412551

1   A ball is dropped from a height of 10 m above the ground. Which of the following best explains what happens to the energy as the ball falls?

   A   The ball gains kinetic energy from its height above Earth.

   B   The ball–Earth system transforms potential energy to kinetic energy.

   C   The Earth–ball system increases gravitational potential energy of the ball.

   D   The gravitational potential energy of the ball is transformed into kinetic energy.

2   A force of 25 N is applied to a box to shift it a distance of 4.5 m across the floor. The floor applies a friction force on the box of 5.0 N. How much kinetic energy is gained by the box?

   A   135 J

   B   113 J

   C   90 J

   D   23 J

3   What is the maximum kinetic energy associated with a 3.0 kg mass when it is thrown downwards at $1.8\,\text{m s}^{-1}$ from a height of 8.4 m above the ground?

   A   257 J

   B   252 J

   C   250 J

   D   247 J

4   A 400 kg motorbike is travelling at $15\,\text{m s}^{-1}$ before accelerating uniformly to $40\,\text{m s}^{-1}$ in 10.2 s. How much work was done by the net force on the motorbike?

   A 9.8×10^{2} J

   B 1.0×10^{4} J

   C 2.8×10^{4} J

   D 5.5×10^{4} J

5   Which of the following is used to explain the law of conservation of energy?

   A   The law of inertia

   B   Newton's first law

   C   Newton's second law

   D   Newton's third law

6 A forwards force on a bicycle and rider is opposed by a combined friction force of 50 N. The speed of the cyclist changes from $4.0\,m\,s^{-1}$ to $10\,m\,s^{-1}$ in 3.0 s and the distance travelled is 24 m. Calculate the total work done on the rider, given that the combined mass of bicycle and rider is 70 kg.

    A 4.6 kJ

    B 3.4 kJ

    C 2.9 kJ

    D 0.72 kJ

7 A groundskeeper pushes down on a 700 kg roller at an angle of 60° to the horizontal with a force of 1.2 kN. The roller slides along the turf for 25 m. How much work is done by the groundskeeper?

    A 15 kJ

    B 30 kJ

    C 0.15 MJ

    D 3.0 MJ

8 How much energy is needed to compress a spring of stiffness $40\,N\,m^{-1}$ by 20 cm?

    A 8.0 kJ

    B 8.0 J

    C 1.6 J

    D 0.80 J

9 What is the maximum amount of energy that can be transformed when a 5.0 kg mass falls into a 4.0 m well from 15 m above the ground?

    A 931 J

    B 735 J

    C 539 J

    D 196 J

10 Which of the following statements is true?

    A The conservation of kinetic energy is true for all collisions.

    B The conservation of kinetic energy and the conservation of momentum are true for elastic collisions only.

    C The conservation of kinetic energy and the conservation of momentum are true for inelastic collisions only.

    D Kinetic energy and momentum are never fully conserved in collisions.

11 A winch pulls a 500 kg boat up a boat ramp, which is 14.7 m long and inclined at an angle of 25° to the horizontal. The winch applies a force that is parallel to the ramp. The ramp applies 2.0 kN of friction to the boat. The winch does 8.3 kJ of work. Calculate the force applied by the winch.

_____

_____

_____

_____

12 Figure 15.4.1 shows the relationship between force and compression of a spring that obeys Hooke's law.

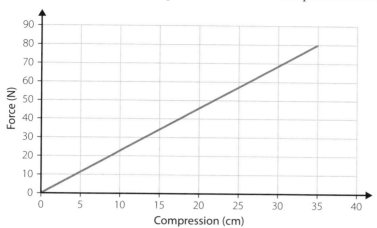

FIGURE 15.4.1

a Find the spring constant.

_____

_____

_____

b Calculate the energy needed to compress the spring from 10 cm to 35 cm.

_____

_____

_____

c A 55 g mass slides along a frictionless surface towards the spring at a speed of $2.3\,\mathrm{m\,s^{-1}}$. By how much does the spring compress? Show your working.

_____

_____

_____

13 A 3.0 kg mass is launched vertically at a speed of $30\,\mathrm{m\,s^{-1}}$ from a height of 20 m. Calculate the kinetic energy when the mass is:

a 15 m above its launch position

_____

_____

_____

b 15 m below its launch position.

_____

_____

_____

14 A 5.6 kg mass travelling to the left at 3.7 m s^{-1} collides with a 6.9 kg mass moving to the left at 2.3 m s^{-1}. They stick together and move off to the left at a speed, v.

    a Find the final velocity, v, of the combined mass.

_____

_____

_____

    b Use calculations to decide whether the collision is elastic or inelastic.

_____

_____

_____

LEARNING

Summary

- All waves transfer energy from one place to another.
- The intensity of a wave is the rate of energy transfer.
- The amplitude of a wave is the maximum displacement of a portion of the wave.
- Mechanical waves require a medium through which to transfer the energy. The medium itself is not transferred, but rather oscillates in place.
- A wavefront is a surface joining all points that are reached at the same instant by a propagating wave.
- The ray model uses a line called a ray that is drawn at right angles to a wavefront to show the direction a wave is travelling.
- A pulse consists of a single wavefront; a continuous wave consists of multiple wavefronts repeating in a regular pattern.
- Transverse waves have oscillations that are perpendicular to the direction of wave travel; longitudinal waves consist of oscillations that are in the direction of wave travel.
- A graph of the displacement of a single particle of the medium as a function of time shows the amplitude (A) and period (T) of a wave. It can also be used to calculate the frequency (f).
- A graph of the displacement vs position of a wave gives a snapshot in time of the position of many particles in the medium and can be used to find the amplitude and wavelength (λ) of the wave.
- The wave velocity is the speed of energy transfer and is dependent on the properties of the medium. Different media have different wave velocities.
- The wave velocity is calculated by $v = f\lambda = \dfrac{\lambda}{T}$.
- At a hard boundary, a wave is reflected in an inverted position; at a soft boundary, a wave is reflected in an upright position. In both cases the transmitted wave is unchanged.
- When a wave is reflected at a boundary, it has an angle of reflection equal to the angle of incidence.
- When a wave is transmitted into a different medium, it is refracted and its velocity, wavelength and angle of travel change. Its frequency is unaffected.
- When a wave is incident on an obstacle, it is diffracted around the obstacle.
- The principle of superposition states that when two waves pass through the same point at the same time, they will produce a resultant wave with an amplitude that is equal to the sum of the amplitude of the individual waves.
- Overlapping waves that are in phase will constructively interfere; waves that are out of phase will destructively interfere.
- Standing waves form when continuous waves of specific frequencies travelling in a medium with boundaries interfere to produce waves that do not appear to travel.
- Nodes are points on standing waves that always destructively interfere, and antinodes are points on standing waves that always constructively interfere.

16.1 Mechanical waves

Mechanical waves are waves that require a medium through which to propagate. The wave does not transfer the medium itself, just the energy. Particles in the medium oscillate around their rest position in response to the energy passing through them. A wavefront is a surface that joins all points in space that are reached at the same instant by a propagating wave, while a ray is a line drawn to indicate the direction of travel of the wave.

If a wave consists of a single wavefront, it is called a pulse; if it consists of multiple wavefronts repeating in a regular pattern, it is termed a continuous wave. A transverse wave is a wave in which the displacement of particles is perpendicular to the motion of a wave, and which consists of crests and troughs. A longitudinal wave is one in which the displacement of the particles is in the direction of wave motion; it consists of a series of compressions and rarefactions.

QUESTIONS

1   Draw a diagram of a spreading circular wave. Include multiple wavefronts and a ray indicating the direction of travel of the wave.

2 Draw a diagram of a continuous transverse wave travelling on a string. Indicate the direction of the oscillation creating the wave, the direction of energy transfer, a crest, a trough and the direction of motion of an individual particle.

3 Draw a diagram of a continuous longitudinal wave travelling on a slinky, and indicate the direction of the oscillation creating the wave, the direction of energy transfer, a compression, a rarefaction and the direction of motion of an individual particle.

16.2 | Wave features

Waves on strings, surface waves on water and seismic S waves are all examples of transverse waves; sound waves and seismic P waves are examples of longitudinal waves. Each wave consists of the motion of particles in their medium.

There are two sinusoidal graphs that describe the features of a wave. A graph of the displacement of a single particle of the medium as a function of time shows the amplitude (A) and period (T) of a wave. From this, the frequency (f) of the wave can be found: $f = \dfrac{1}{T}$. A graph of displacement vs position gives a snapshot in time of the position of many particles in the medium, and can be used to find the amplitude and wavelength (λ) of the wave. The wave velocity (v) can be found by the formula $v = f\lambda = \dfrac{\lambda}{T}$. The wave velocity is dependent on the medium itself and can vary considerably.

QUESTIONS

1 Classify each of the following wave types as transverse or longitudinal.

a Vibration on a string

_____

b Sound waves

_____

c Seismic S waves

_____

d Seismic P waves

_____

e Surface waves on water

_____

9780170412551

2 Analyse the graphs of continuous waves in Figure 16.2.1 and determine the following values.

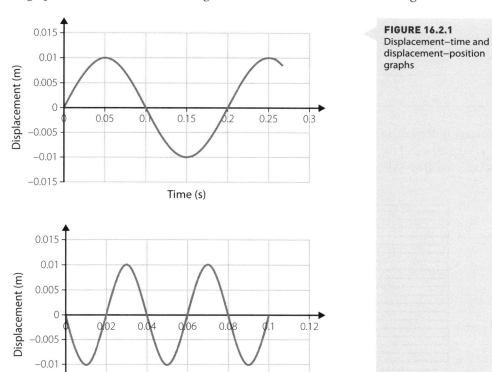

FIGURE 16.2.1
Displacement–time and displacement–position graphs

a The amplitude

_____

_____

b The wavelength

_____

_____

c The period

_____

_____

d The frequency

_____

_____

e The wave velocity

_____

_____

3 The time taken for successive ocean waves to pass under a boat is approximately 4.2 s and the distance between them is 21 m. Determine the speed at which the ocean waves are travelling.

_____

_____

_____

4 Analyse the following diagram of seismic waves arriving at a seismograph, to determine the distance between the source of the earthquake and the seismograph if it is known that S-waves travel at a speed of $1.3\,\text{km}\,\text{s}^{-1}$ and P waves travel at a speed of $7.4\,\text{km}\,\text{s}^{-1}$.

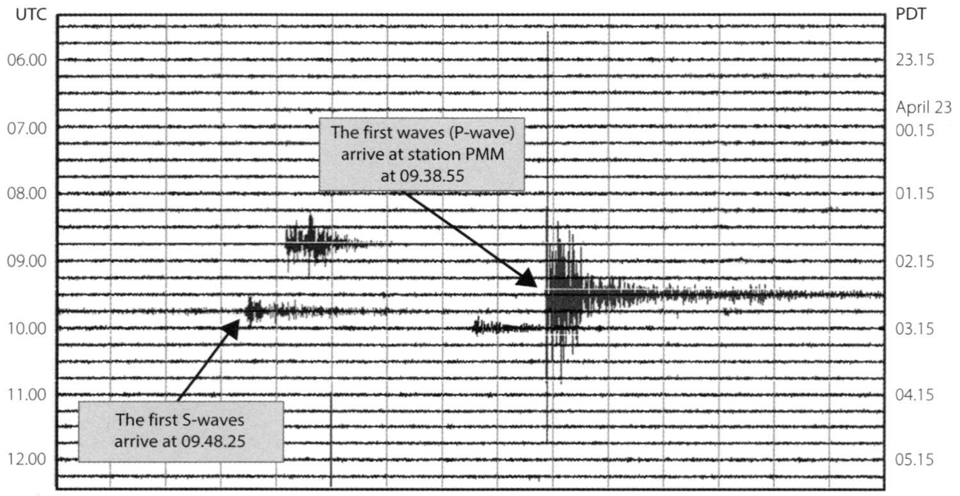

UTC / PDT columns:
- 06.00 / 23.15
- 07.00 / April 23 00.15
- 08.00 / 01.15
- 09.00 / 02.15
- 10.00 / 03.15
- 11.00 / 04.15
- 12.00 / 05.15

The first waves (P-wave) arrive at station PMM at 09.38.55

The first S-waves arrive at 09.48.25

FIGURE 16.2.2 Seismic S and P waves recorded by seismograph

_____

_____

_____

_____

_____

_____

_____

_____

_____

_____

_____

_____

_____

_____

9780170412551

16.3 | Reflection

Reflection occurs when a wave travelling in a medium is incident on a boundary with a different medium. If it strikes a fixed boundary, the wave is reflected in an inverted position relative to the incident wave. If it strikes a free boundary, it is reflected in the upright position. The wave can also be transmitted into the new medium. If the density of the new medium is higher than that of the first medium, the wave will be upright and it will slow down; if the new medium has a lower density, the wave will be upright and it will speed up. When a wave is incident on a boundary at an incident angle greater than zero, it will be reflected on the other side of a normal to the boundary at the point of incidence at an angle equal to the angle of incidence. This gives rise to phenomena such as total internal reflection, reverberations and echoes.

QUESTIONS

1 A wave pulse travelling along a string is incident upon a boundary. Draw the orientation of the reflected wave if the boundary is:

   a   fixed

   b   free to move.

2 A wave pulse travelling on a string is incident on a boundary with a string of greater density. Draw the orientation of the reflected and transmitted pulses.

3 Draw a diagram that represents a wave striking a boundary with an angle of incidence of 30°. Include the following features:

  a incident ray

  b angle of incidence

  c normal

  d reflected ray

  e angle of reflection.

9780170412551

16.4 Refraction and diffraction

When a wave is transmitted into a different medium, the difference in the elastic potential and/or mass density of the two media results in a change of velocity and wavelength. If the wave is incident at any angle other than 90°, it will also be bent as it is transmitted. This results in many optical illusions including a change in the apparent position of an underwater object and the flattening of the Sun as it nears the horizon.

When a wave is incident upon an object or boundary, it will diffract around the obstacle and move into the region behind it. The degree to which this occurs is greatest when the wavelength of the incoming wave is greater than the size of the obstacle. This is most evident at the entrance to a harbour, where incoming plane waves diffract in circular patterns around the sea walls.

QUESTIONS

1 Complete the following sentences using the word options provided.

| refraction | transmission | mass density | medium | wavelength |
| speed | elastic | direction | waves | |

The refraction of _____ involves a change in the _____ of the waves as they pass from

one _____ to another. _____, or bending of the path of the waves, is accompanied by changes in

the _____ and _____ of the waves. This is due to the new medium having a

different _____ property and/or _____ that affects the rate of _____ of the wave energy.

2 Draw an image of the wavefronts on both sides of a boundary between two media if the wave is incident at right angles to a boundary. Include both the incident and refracted rays and an indication of the wavelengths of the incident and refracted waves.

3 Draw an image of the wavefronts on both sides of a boundary between two media if the wave is incident at an angle other than 90° to the boundary. Include both the incident and refracted rays, the incident and refracted angles and an indication of the wavelengths of the incident and refracted waves.

4 Write a short paragraph describing the concept of diffraction using the following words.

diffraction	obstacle	gap	bend	behind
amount	dependent	incident	greatest	wavelength

_____

_____

_____

_____

_____

_____

_____

_____

_____

_____

_____

5 On the images in Figure 16.4.1, draw in the diffraction that would be expected for each scenario given the relative sizes of the incident wavelengths and gap widths.

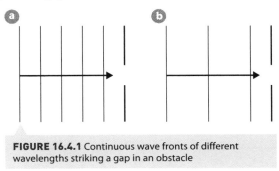

FIGURE 16.4.1 Continuous wave fronts of different wavelengths striking a gap in an obstacle

16.5 | The principle of superposition

The principle of superposition states that when two waves travelling through the same medium overlap in the same place at the same time, the waves will interfere with each other to produce a resulting wave that has an amplitude equal to the sum of their individual amplitudes. If the coherent waves are entirely in phase, constructive interference will occur to produce an increased maximum, while if the two coherent waves are entirely out of phase, destructive interference will occur that will produce an amplitude of zero.

If continuous waves are travelling in a medium that has boundaries, the reflection of these waves at the boundaries will cause the waves to interfere according to the principle of superposition. At certain frequencies and wavelengths, these waves will form standing wave patterns that are waves appearing to oscillate in place. These waves have nodes, which are points where destructive interference always occurs, and antinodes, which are points where constructive interference always occurs. Musical instruments make use of this phenomenon to produce their characteristic tones.

QUESTIONS

1 In Figure 16.5.1 and Figure 16.5.2, two wave pulses are travelling towards each other on a string. Classify the pulses in each scenario as being in phase or out of phase. Draw a diagram of the displacement of the string when the two pulses overlap and classify each scenario as either constructive interference or destructive interference.

a

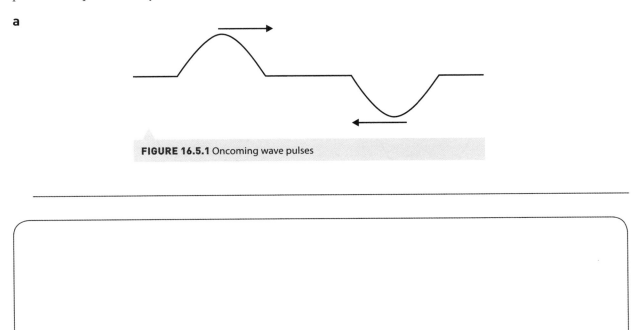

FIGURE 16.5.1 Oncoming wave pulses

b

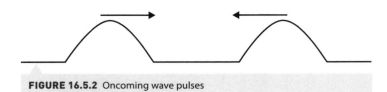

FIGURE 16.5.2 Oncoming wave pulses

---

2 The diagram below shows two waves that are passing along a string at the same time. Apply the principle of superposition to draw in the wave that would appear on the string because of their interaction.

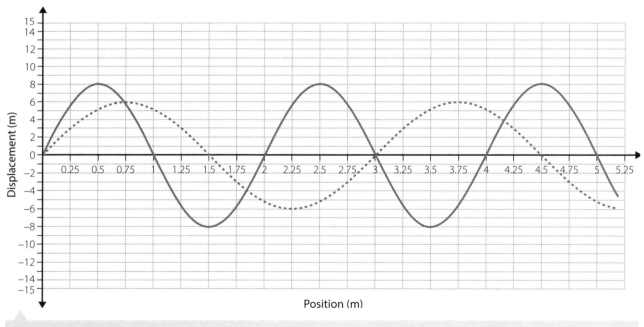

FIGURE 16.5.3 Two overlapping waves

3 On the diagram of a standing wave on a string shown in Figure 16.5.4, label the nodes and antinodes and explain what motion the particles of the string at these positions will undergo.

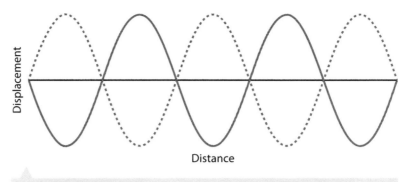

FIGURE 16.5.4 A standing wave

_____

_____

_____

_____

EVALUATION

1 The intensity of a wave is dependent upon:

    A the amplitude of the wave.

    B the wavelength of the wave.

    C the frequency of the wave.

    D the period of the wave.

2 The direction of particle motion in a medium containing a transverse wave is:

    A parallel to the velocity of the wave.

    B perpendicular to the velocity of the wave.

    C at an acute angle to the velocity of the wave.

    D at an obtuse angle to the velocity of the wave.

3 Which of the following features can be deduced from a displacement–distance graph of a wave?

    A Period

    B Frequency

    C Intensity

    D Wavelength

4 If a wave pulse travelling in a light string is incident upon a boundary with a heavy string, then the transmitted wave pulse will be:

    A upright and diminished.

    B upright and amplified.

    C inverted and diminished.

    D inverted and amplified.

5 If a wave strikes a surface at an angle of 15° to the surface, then the angle of reflection of the wave is:

    A 15°.

    B 90°.

    C 75°.

    D 0°.

9780170412551

6 When two wave pulses that are out of phase meet at a point, their interaction results in:

   A constructive interference.

   B destructive interference.

   C a standing wave.

   D total internal reflection.

7 What is required for a mechanical wave to propagate?

_____

8 What is the name of the relationship given to two wave pulses whose amplitude is orientated in the same direction?

_____

9 What is the name given to the points on a standing wave that do not move?

_____

10 Draw a diagram of a circular wave expanding from a point source. Include the wave fronts and six rays indicating the direction of travel.

11 Compare the features of a longitudinal wave and a transverse wave.

_____

_____

_____

_____

_____

12 Analyse the following diagrams and sketch what would be seen 10.0 s after what is shown in the first instance.

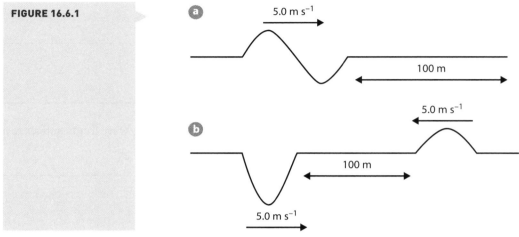

FIGURE 16.6.1

13 If the wavelength of sound produced by a clarinet is 0.8575 m and its frequency is 400.0 Hz, determine:

a the period of the sound wave

_____

_____

_____

_____

b the velocity of the sound wave.

_____

_____

_____

14 Analyse the graphs of a wave on a string shown in Figures 16.6.2 and 16.6.3 and determine:

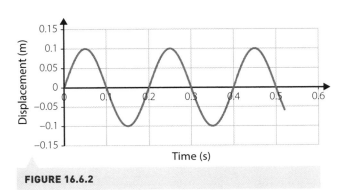

FIGURE 16.6.2

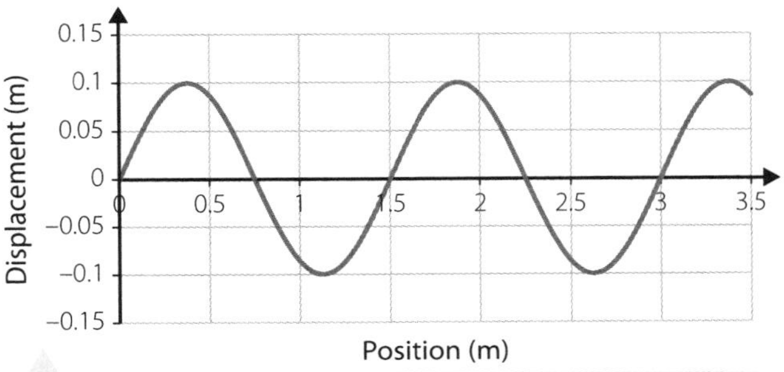

FIGURE 16.6.3

a the amplitude of the wave

_____

b the period of the wave

_____

c the frequency of the wave

_____

d the wavelength of the wave

_____

e the velocity of the wave.

_____

15 Explain how standing waves are formed on stringed instruments.

_____

_____

_____

_____

_____

_____

16 How is sonar, as used by bats, dolphins and whales, similar to the principles used by radar?

_____

_____

_____

_____

_____

_____

17 Explain how we know so much about the internal structure of Earth.

_____

_____

_____

_____

_____

_____

_____

9780170412551

17 Sound

Summary

▶ The natural frequency of an object is the frequency with which it will vibrate if it is displaced from its equilibrium position and then left to vibrate by itself.

▶ A forced vibration occurs when one vibrating object makes another object vibrate at the same frequency.

▶ When the forced vibration coincides with the natural frequency of an object, resonance occurs that transfers energy very efficiently.

▶ The vibrational modes or harmonics that form on a string are dependent on the length of the string and the elastic properties of the string.

▶ The wavelength of the harmonics can be calculated as $\ell = n\,\dfrac{\lambda_n}{2}$, where ℓ is the length of the string, n is the harmonic number and λ is the wavelength.

▶ The frequency of the harmonics can be calculated as $f_n = \dfrac{v}{\lambda_n} = n\dfrac{v}{2\ell} = nf_1$ where f_n is the frequency and v is the velocity of the wave.

▶ An open air column is a tube that has both of its ends open to the surroundings.

▶ The wavelength of the harmonics of an open air column can be calculated as $\ell = n\,\dfrac{\lambda_n}{2}$.

▶ The frequency of the harmonics of an open air column can be calculated as $f_n = \dfrac{v}{\lambda_n} = n\dfrac{v}{2\ell} = nf_1$.

▶ A closed air column is a pipe that has one end closed to the surroundings.

▶ The wavelength of the harmonics of a closed air column can be calculated as $\ell = (2n-1)\,\dfrac{\lambda_n}{2}$.

▶ The frequency of the harmonics of a closed air column can be calculated as $f_n = \dfrac{v}{\lambda_n} = (2n-1)\dfrac{v}{2\ell} = (2n-1)\,f_1$.

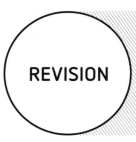

17.1 Resonance

Any object is capable of vibrating after being displaced. The natural frequency of an object is the frequency with which it will vibrate if it is displaced from its equilibrium position and then left to vibrate by itself. A forced vibration occurs when one vibrating object makes another object vibrate at the same frequency. When the forced vibration coincides with the natural frequency of an object, resonance occurs that transfers energy very efficiently.

QUESTIONS

1 Create a glossary containing the following terms.

   a Natural frequency

   b Forced vibration

   c Resonance

   d Driving frequency

   _____

   _____

   _____

   _____

   _____

   _____

   _____

   _____

2 Complete these sentences about resonance by inserting the words provided below.

natural	frequency	resonance	amplitude
efficiency	energy	resonating	

   • _____ will only occur when the driving _____ matches the _____ frequency.

   • The _____ of the vibration of the _____ object will increase dramatically.

   • When an object is resonating, _____ is being transferred with maximum _____ from the driving oscillator to the receiving oscillator.

17.2 Vibrating strings

Every object has a number of standing wave patterns that can form upon its length. In a string, these vibrational modes or harmonics depend on the length and elastic properties of the string itself. These standing waves have distinct wavelengths and frequencies, as well as patterns of nodes and antinodes. The relationship between the string length and the wavelength of the harmonic can be shown as $\ell = n\,\dfrac{\lambda_n}{2}$, where ℓ is the length of the string, n is the harmonic number and λ is the wavelength. The frequency of these waves can be calculated as $f_n = \dfrac{v}{\lambda_n} = n\dfrac{v}{2\ell} = nf_1$, where f_1 is the resonant frequency of the string, f_n is the frequency of the standing wave and v is the velocity of the wave.

QUESTIONS

1 Draw a diagram of the first three standing wave patterns that will occur on a vibrating string that is 1.2 m long, and determine the wavelength of each.

2 A piano string produces middle C, which has a frequency of 256 Hz. What are the frequencies of the second, third and fourth harmonics of this piano string?

3 A wire is stretched between two stationary supports that are 45 cm apart.

a Determine the wavelength of the fundamental mode of vibration.

b Calculate the fundamental frequency of the wire if waves travel along it at $200\,\text{m s}^{-1}$.

_____

_____

_____

c Determine the distance between nodes in the third harmonic that forms on the wire.

_____

_____

_____

17.3 | Air columns

The resonant modes or harmonics that form in air columns are dependent upon the length of the air column, the velocity of sound in the air column, and whether the column has a closed end or not. The compressions and rarefactions that travel in an air column are reflected from the walls of the pipe. When a compression or rarefaction is reflected from the end of an open pipe, the pressure wave is inverted; a compression is reflected as a rarefaction, and a rarefaction is reflected as a compression. However, when they are reflected at the end of a closed pipe, compressions are reflected as compressions, and rarefactions as rarefactions.

The relationship between the length of an open pipe and wavelength of the harmonics is $\ell = n\,\dfrac{\lambda_n}{2}$, with the frequency calculated as $f_n = \dfrac{v}{\lambda_n} = n\dfrac{v}{2\ell} = nf_1$. The relationship between the length of a closed pipe and the wavelength of its harmonics is $\ell = (2n-1)\,\dfrac{\lambda_n}{4}$, with their frequencies calculated as $f_n = \dfrac{v}{\lambda_n} = (2n-1)\dfrac{v}{4\ell} = (2n-1)\,f_1$.

In each case, ℓ is the length of the pipe, λ_n is the wavelength of the standing wave, n is the harmonic number, f_n is the frequency of the standing wave, v is the velocity of air in the pipe and f_1 is the fundamental frequency of the pipe.

QUESTIONS

1 Draw a diagram of both the particle displacement and the pressure variation of the second harmonic that will form in an open air column and a closed air column, and determine the frequency of each, if the velocity of air in the column is $340\,\text{m s}^{-1}$ and the columns are 2.2 m long.

9780170412551

2 The fundamental frequency of a 1.5 m long open air column is 115 Hz.

a Determine the frequency of the next three harmonics.

_____

_____

b Calculate the velocity of air in the column.

_____

_____

c Calculate the wavelength of the second harmonic.

_____

_____

d Determine whether or not a standing wave of frequency 920 Hz will form in the column.

_____

_____

_____

3 An experiment is carried out by blowing across the top of the 32 cm bottle shown below.

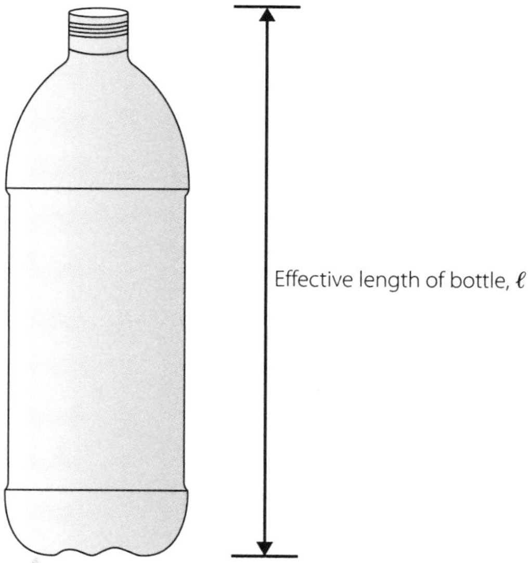

Effective length of bottle, ℓ

FIGURE 17.3.1

a Determine the wavelength of the fundamental frequency that should be created in the bottle.

_____

_____

b Determine the fundamental frequency of the bottle assuming the velocity of sound in air is $340 \, \text{m s}^{-1}$.

_____

_____

c If a tone of frequency 320 Hz is emitted when air is blown across, determine the percentage difference between the predicted and actual frequencies.

_____

_____

d Suggest reasons why the two frequencies differ.

_____

_____

_____

_____

_____

9780170412551

1 A forced vibration occurs when:

A an object is displaced then left to vibrate.

B an object vibrates due to the vibration of another object.

C the vibration of one object coincides with the resonant frequency of another.

D a standing wave forms in an object.

2 The resonant frequency of an object is:

A the frequency at which is vibrating.

B the frequency that it will vibrate naturally.

C the frequency at which it will vibrate due to another object.

D the third harmonic standing wave mode.

3 When an object is resonating:

A energy is being transferred with maximum efficiency.

B its amplitude will decay quickly.

C it will emit tones of varying pitches.

D it is causing another object to vibrate.

4 What is the number of nodes that exist on a stretched string vibrating at its third harmonic mode?

A 2

B 3

C 4

D 5

5 What is the number of antinodes existing on a closed air column vibrating at its second harmonic mode?

A 1

B 2

C 3

D 4

6 What is the term given to the lowest possible frequency produced by a system?

_____

7 What is the term given to the point on a standing wave that undergoes complete destructive interference at all times?

_____

8 Calculate the wavelength of the first three harmonic modes that can form on a string that is 1.3 m long.

_____

_____

_____

_____

_____

9 A wire is stretched between two stationary supports that are 1.2 m apart.

    a Determine the wavelength of the fundamental mode of vibration.

_____

_____

    b Determine the fundamental frequency of the wire if waves travel along it at $230.0\,\mathrm{m\,s^{-1}}$.

_____

_____

    c Calculate the distance between nodes in the fourth harmonic that forms on the wire.

_____

_____

_____

9780170412551

10 The fundamental frequency of a 1.8 m long open air column is 140 Hz.

a Determine the frequency of the next three harmonics.

_____

_____

b Calculate the velocity of air in the column.

_____

_____

c Calculate the wavelength of the third harmonic.

_____

_____

d Determine whether or not a standing wave of frequency 1680 Hz will form in the column.

_____

_____

_____

11 If the velocity inside a 2.60 m closed pipe is 343 m s^{-1}, determine the:

a wavelengths of the first three harmonic modes that will form in the pipe

_____

_____

b distance between adjacent antinodes that will form in the fourth harmonic mode

_____

_____

c fundamental frequency of the pipe.

_____

_____

_____

_____

_____

_____

_____

12 If the velocity of sound inside two pipes is such that a 1.3 m closed pipe produces a harmonic wave of frequency 336.5 Hz at the same time that a 1.4 m open pipe produces a harmonic wave of frequency 250.0 Hz, determine the velocity of sound in the pipes.

_____

_____

_____

_____

_____

_____

13 Explain why an earthquake can cause a building to oscillate violently even though the amplitude of the earthquake may be much lower than that of the building.

_____

_____

_____

_____

_____

_____

14 Even with your eyes closed you could tell the difference between a clarinet and a saxophone when they are played. Explain what is different about the sound produced by different instruments.

_____

_____

_____

_____

_____

_____

9780170412551

18 Light

LEARNING

Summary

- Light can be modelled using the ray model, the wave model or the particle model.
- A light wave is a form of electromagnetic radiation that interacts with matter as a three-dimensional transverse wave.
- Luminous objects emit light, while non-luminous objects reflect light.
- The speed of light is dependent upon the medium through which it is travelling.
- The speed of light in a vacuum or air is $c = 3.0 \times 10^8 \, \text{m s}^{-1}$.
- The intensity of light at a point is inversely proportional to the square of the distance from the source.
- The fact that light can be polarised gives proof that light is a transverse wave.
- Regular reflection occurs off a smooth reflective surface while diffuse reflection occurs off a rough surface.
- The law of reflection states that the angle of reflection is equal to the angle of incidence.
- The image formed by a reflection is called a virtual image because it appears to form behind the reflecting surface.
- The magnification of an image is equal to the ratio of the size of the image and the size of the object: $M = \dfrac{h_i}{h_o}$, where M is the magnification, h_i is the height of the image and h_o is the height of the object.
- The refractive index, n, gives an indication of the refrangibility of a medium.
- The amount of refraction of an incident light wave is determined by the ratio of the refractive index of the two media: $\dfrac{\sin i}{\sin R} = \dfrac{n_R}{n_i}$.
- The wavelength and speed of incident light will change when it is refracted: $\dfrac{\sin i}{\sin R} = \dfrac{\lambda_i}{\lambda_R} = \dfrac{v_i}{v_R} = \dfrac{n_R}{n_i}$.
- Total internal reflection occurs when all incident light is reflected and not refracted at the boundary with a new medium. This occurs when the angle of incidence is greater than the critical angle: $i_c = \sin^{-1}\left(\dfrac{n_2}{n_1}\right)$.
- Chromatic dispersion occurs because different wavelengths of light are refracted by differing amounts.
- Convex lenses refract light towards the principle axis to form a real image.
- Concave lenses will refract light away from the principle axis to form a virtual image.
- Diffraction of light occurs when light is incident upon an aperture or object with a width comparable to the wavelength of light.
- A diffraction pattern will be created with a central bright spot separated from less intense bright spots by dark spots.

- When light is incident upon two slits, it will form a distinct pattern of bright and dark spots caused by the constructive and destructive interference of the two diffraction patterns.
- A bright spot occurs at points where constructive interference is occurring when the path length difference is $n\lambda$.
- A dark spot occurs at points where destructive interference is occurring when the path length difference is $(2n-1)\dfrac{\lambda}{2}$.

9780170412551

18.1 Models of light

In the ray model of light, light is described as travelling in straight lines from its source and changes direction when it interacts with a material. In the wave model, light is treated as a wave that can travel both through a medium and a vacuum. In the particle model, light is treated as consisting of particles called photons.

The electromagnetic wave model of light describes light as a form of electromagnetic wave that acts as a three-dimensional transverse wave in its interactions with matter. The propagation of waves can be described by the ray model, which describes waves as travelling in straight lines from a source. The source can be either luminous, if it produces light, or non-luminous, if it reflects light. The speed of light is dependent upon the medium through which it travels. The speed of light in air or a vacuum is $3.0 \times 10^8\,\mathrm{m\,s^{-1}}$. The intensity of light is calculated by the rate of energy passing through a set area, and can be shown to be inversely proportional to the square of the distance from the source $I \propto \dfrac{1}{r^2}$.

QUESTIONS

1 Complete the following statement about the nature of light by inserting the given terms.

matter	experiments	light	particle
models	duality	both	wave

In some experiments, _____ seems to travel as a _____, but to interact with _____ as a _____. These _____ cannot be explained without the use of _____ the wave and particle _____ together. Scientists call this need for these two apparently quite different models the wave–particle _____.

2 Decide which model of light (the ray model, wave model or particle model) is most useful to describe the following scenarios.

a Light passing through a lens

_____

b Light bending around an object

_____

c Light being absorbed by an atom

_____

d The path of light through space

_____

e The absorption of light on a solar panel

_____

f Interference patterns of light

_____

3 Students measured the intensity of light at various distances from a source and tabulated their results in Table 18.1.1 below:

TABLE 18.1.1 Students' results

DISTANCE FROM SOURCE (m)	INTENSITY OF LIGHT (W m^{-1})
1	28
2	7
3	3.1
4	1.75
5	1.12
6	0.78
7	0.57
8	0.44
9	0.35
10	0.28

a Graph the data to show that it obeys the inverse square law: $I \propto \dfrac{1}{r^2}$.

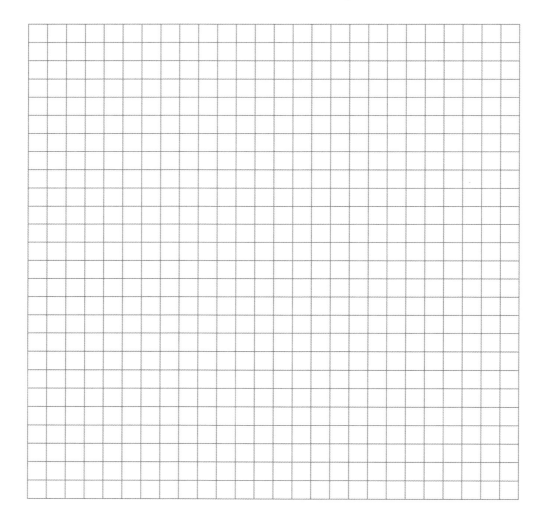

b State a formula of best fit that shows this relationship.

_____

9780170412551

4 Present a diagram showing wavefronts and rays of light propagating in the following scenarios.

a Circular waves emanating from a point source

b Parallel waves emanating from a surface

c Circular waves appearing as parallel waves when observed at a great distance

18.2 Polarisation and the transverse model

The fact that light can be polarised shows that it is a transverse wave. Polarisation occurs when the oscillations of the waves are all aligned in one direction. This can be done by passing light first through a polariser, which only transmits light that is orientated in a particular direction, and then through an analyser, which will only transmit light if it is orientated in the direction of the polariser.

QUESTIONS

1 a Draw an annotated diagram of a polariser and an analyser that allows light waves to pass through.

9780170412551

b Draw an annotated diagram of a polariser and an analyser that does not allow light waves to pass through.

18.3 | Reflection of light

The reflection of light obeys the law of reflection of waves that states that the angle of incidence (the angle between the incoming ray and a normal to the surface) is equal to the angle of reflection (the angle between the reflected ray and the same normal). Regular reflection occurs when light is reflected off a smooth surface in a predictable way. Diffuse reflection occurs when light is reflected off a rough surface in all directions. The reflected image that is seen by an observer appears to be behind the surface; this is a virtual image. The magnitude (M) of an image is the ratio of the image height to the object height: $M = \dfrac{h_i}{h_o}$.

QUESTIONS

1 Draw and annotate two diagrams that clearly show:

a specular reflection

b diffuse reflection.

2 Complete the following diagram representing the reflection of an arrow (PO) in a mirror (MM′) by drawing in two rays that reflect at points A and B respectively. Use this diagram to show:

a that the image formed will be virtual

b that the image formed will have a magnification of 1.

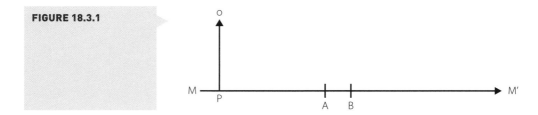

FIGURE 18.3.1

18.4 | Snell's law and the refraction of light

Refraction of light occurs at the boundary between two mediums whenever the angle of incidence is greater than 0°. The refrangibility of a medium is a measure of the amount of refraction that will occur in it, and is described by its refractive index. The refrangibility of a medium also gives an indication of the speed of light in that medium as well as the wavelength. Snell's law of refraction states that $\dfrac{\sin i}{\sin R} = \dfrac{\lambda_i}{\lambda_R} = \dfrac{v_i}{v_R} = \dfrac{n_R}{n_i}$, where n is the refractive index of a medium, i is the angle of incidence, R is the angle of refraction and the subscripts $_i$ and $_R$ indicate incident and refraction, respectively. Total internal reflection occurs at a boundary between two media when all incident light is reflected and none is transmitted. This occurs at incident angles greater than the critical angle $i_c = \sin^{-1}\left(\dfrac{n_2}{n_1}\right)$.

Chromatic dispersion, which is the separation of light into its constituent colours, occurs because different wavelengths are refracted by differing amounts.

Lenses make use of the refraction of light to bend light in predictable and useful ways. A convex lens refracts incident light towards the principle axis, with the formation of a real image on the opposite side of the lens. A concave lens will refract light away from the principle axis, to form a virtual image on the same side of the lens as the object.

QUESTIONS

1 Use the following terms to construct a paragraph regarding the refraction of light.

speed	medium	permitted	refraction
electrical	light	direction	

2 Complete the following ray diagrams to show the direction in which light is bent when the relationship between refractive indices of the two medium is:

a $n_1 < n_2$

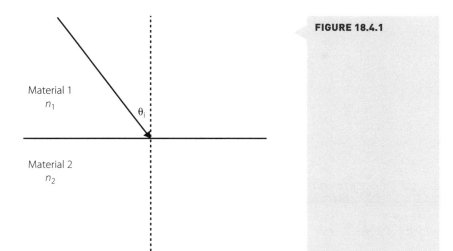

FIGURE 18.4.1

Material 1
n_1

θ_i

Material 2
n_2

b $n_1 > n_2$.

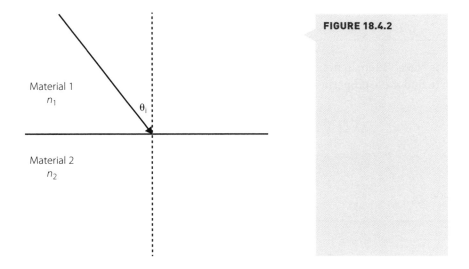

FIGURE 18.4.2

Material 1
n_1

θ_i

Material 2
n_2

3 Light of wavelength 520 nm travels in air ($n_a = 1.00$) before it strikes the interface with water ($n_w = 1.33$) at an angle of 45° to the normal.

a What is the wavelength of the light in water?

_____

_____

_____

b What is the angle of refraction in the water?

_____

_____

_____

c Draw a diagram of the scenario.

d If the light has a velocity of $2.26 \times 10^8 \, \mathrm{m\,s^{-1}}$ when it is in the water, with what velocity must it have been travelling in air?

4 An object 3 cm high is placed 12 cm in front of a converging lens of focal length 6 cm. Use an accurately drawn ray tracing diagram to find these properties of the image.

a Its position

b Its nature

c Its size

d The magnification

_____

_____

_____

_____

_____

_____

_____

_____

_____

_____

5 An object 3 cm high is placed 12 cm in front of a diverging lens of focal length 6 cm. Use an accurately drawn ray tracing diagram to find these properties of the image.

a Its position

b Its nature

c Its size

d The magnification

_____

_____

_____

_____

_____

_____

_____

18.5 | Diffraction

When light is incident upon a narrow aperture or object of similar width to the wavelength of light, it diffracts into a distinctive diffraction pattern consisting of a wide central bright fringe and several less intense bright fringes separated by dark spots. When light is incident upon two narrow apertures, as in Young's double-slit experiment, the interference of the two diffraction patterns formed will form a distinctive pattern of bright fringes indicating that the two patterns are constructively interfering, and dark fringes indicating that destructive interference is occurring. Constructive interference will occur at points when the difference in length between the two paths is equal to $n\lambda$, where n is any integer 0, 1, 2, 3... Destructive interference will occur when the difference in length between the two paths is equal to $(2n-1)\dfrac{\lambda}{2}$, where n is any integer 0, 1, 2, 3...

QUESTIONS

1 Complete the following diagram of the wavefront behaviour that would be observed if plane waves are incident on a single slit as shown below.

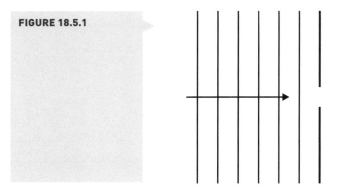

FIGURE 18.5.1

2 Complete the following diagram of the wavefront behaviour that would be observed if plane waves are incident on a double-slit apparatus as shown below. Indicate points that represent waves in phase.

FIGURE 18.5.2

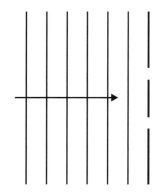

9780170412551

3 Determine whether the points in the following table, which represent distances from individual slits in the double-slit apparatus, would result in constructive or destructive interference when light of 500 nm is incident upon the slits.

DISTANCE FROM SLIT 1 (nm)	DISTANCE FROM SLIT 2 (nm)	CONSTRUCTIVE OR DESTRUCTIVE
500	1000	
500	1250	
500	1300	
1250	1750	
2560	2690	
2560	2810	
2560	3060	

1   The propagation of light is best described using which of the following?

   A The ray model of light

   B The wave model of light

   C The particle model of light

   D The corpuscular model of light

2   Which of the following is an example of a luminous object?

   A A desk top

   B A sheet of paper

   C A desk lamp

   D A pen

3   Which of the following is not a correct description of the direction travelled by light?

   A Light travels in straight lines.

   B Light travels equally in all directions from a source.

   C Light travels at right angles to the oscillations in the electric field.

   D Light travels in the same direction as the oscillations in the magnetic field.

4   Opaque objects are visible from many directions because of:

   A diffuse scattering of light.

   B specular reflection of light.

   C refraction of light.

   D diffraction of light.

5   The refrangibillity of a medium is dependent upon:

   A the colour of the object.

   B the size of the object.

   C the electrical properties of the object.

   D the temperature of the object.

6 Which of the following colours undergoes the greatest degree of refraction?

  A Red

  B Yellow

  C Green

  D Blue

7 What is the term given to the phenomena that light can behave both as a wave and as a particle?

_____

8 In the particle model of light, what is the name given to a particle of light?

_____

9 What is the term given to a light wave that strikes a surface?

_____

10 State the law of reflection.

_____

_____

11 Which properties of a light wave will change due to refraction?

_____

_____

12 If the intensity of a light source at a particular point is measured to be $150\,\mathrm{W\,m^{-2}}$, determine the intensity of the same light source two times further out.

_____

_____

_____

13 Light of wavelength 445 nm travels in water ($n_w=1.33$) before it strikes the interface with air ($n_a=1.00$) at an angle of 30.0° to the normal.

  a Determine the wavelength of the light in air.

_____

_____

_____

  b Determine the angle of refraction in the air.

_____

_____

_____

c If the light has a velocity of $3.0 \times 10^8 \, \text{m s}^{-1}$ when it is in the air, with what velocity must it have been travelling in water?

_____

_____

_____

d Calculate the critical angle for this boundary.

_____

_____

_____

14 An object 5 cm high is placed 10 cm in front of a converging lens of focal length 15 cm. Use an accurately drawn ray tracing diagram to find these properties of the image.

   a Its position

   b Its nature

   c Its size

   d The magnification

_____

_____

_____

_____

_____

15 Explain how the double-slit experiment could be used to determine the wavelength of light that is incident upon it.

_____

_____

_____

_____

_____

16 Explain how light is 'guided' in optical fibres.

_____

_____

_____

_____

_____

PHYSICS UNITS 1 & 2

MULTIPLE-CHOICE QUESTIONS

Question 1

Thermal equilibrium exists between two substances when:

A the temperature of both substances in contact is the same.

B both substances are in contact.

C the potential energy of the cooler substance is given to the hotter substance.

D the heat from the colder substance is transferred to the hotter substance.

Question 2

23°C on the Kelvin scale is:

A 250°C.

B 250 K.

C 296 K.

D 296°C.

Question 3

How much energy is required to heat a 5.0 kg bag of ice from −4°C to 15°C?

Specific heat capacity of solid water (ice) = 2100 J kg^{-1} K^{-1}

Specific heat capacity of liquid water = 4200 J kg^{-1} K^{-1}

Latent heat of fusion = 334 kJ kg^{-1}

A 0.30 MJ

B 0.36 MJ

C 0.41 MJ

D 2.0 MJ

Question 4

How much energy is needed to raise the temperature of 200 mL of milk from 5°C to 50°C? Assume that the specific heat capacity of milk is 4010 J kg^{-1} K^{-1}.

A 3600 J

B 3.6×10^4 J

C 3.6×10^6 J

D 3.6×10^7 J

Question 5

Which of the following best describes the structure of an atom?

A A tightly bonded collection of negatively charged protons and neutrons (no charge) in the nucleus, which are surrounded by a cloud of small, positively charged electrons.

B A tightly bonded collection of positively charged positrons and neutrons (no charge) in the nucleus, which are surrounded by a cloud of small, negatively charged electrons.

C A tightly bonded collection of positively charged protons and neutrons (no charge) in the nucleus, which are surrounded by a cloud of small, negatively charged electrons.

D A tightly bonded collection of negatively charged electrons and neutrons (no charge) in the nucleus, which are surrounded by a cloud of small, positively charged positrons.

Question 6

Which of the following emissions have the least ionising effect?

A Alpha particles

B Beta particles

C Gamma rays

D None of the above – they all have strong ionising properties.

Question 7

Terbium, Tb–148 has 83 neutrons. It decays by positron emission to an isotope of gadolinium, Gd. The gadolinium nuclide then alpha decays to samarium, Sm-144. What are the symbols for the nuclide of terbium-148 and samarium-144 in this decay series?

A $^{83}_{65}\text{Tb}$; $^{79}_{62}\text{Sm}$

B $^{148}_{83}\text{Tb}$; $^{144}_{79}\text{Sm}$

C $^{65}_{148}\text{Tb}$; $^{62}_{144}\text{Sm}$

D $^{148}_{65}\text{Tb}$; $^{144}_{62}\text{Sm}$

Question 8

In the neutron bombardment of aluminium, a new isotope of aluminium is formed. What is the other product of the reaction?

$$^{1}_{0}\text{n} + {}^{27}_{13}\text{Al} \rightarrow {}^{28}_{13}\text{Al} + ?$$

A An alpha particle

B A beta particle

C A gamma ray

D None of the above

Question 9

Electric potential is best described as:

A the rate of energy transfer between points in a circuit.

B the flow of voltage.

C the amount of energy available per unit charge.

D the number of cells in a battery.

Question 10

Which of the following is a CORRECT statement about circuits?

A In a series circuit, the current is shared between components; the potential difference is the same across all components.

B In a series circuit, the current is the same in all components; the potential difference is the same across all components.

C In a parallel circuit, the current is shared between components; the potential difference is the same across all components.

D In a parallel circuit, the current is the same in all components; the potential difference is the same across all components.

Question 11

Use the graph to find the distance travelled by the object in the first 4.0 s.

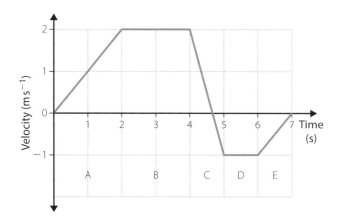

A 2.0 m

B 4.0 m

C 6.0 m

D 8.0 m

Question 12

A driver heading north wishes to cover 850 km in a period of 10 hours. After travelling for 5 hours, they have averaged 90 km h^{-1}. What average velocity will be required over the second 5 hours to reach this goal?

A 75 km h^{-1}

B 80 km h^{-1}

C 85 km h^{-1}

D 90 km h^{-1}

Question 13

Hooke's law is often written as $F = kx$. What do the symbols, F, k and x represent?

A F represents the energy applied to the spring; k represents the energy stored in the spring; x represents the extension of the spring.

B F represents the energy applied to the spring; k represents the stiffness; x represents the length of the spring.

C F represents the force applied to the spring; k represents the energy stored in the spring; x represents the extension of the spring.

D F represents the force applied to the spring; k represents the spring constant; x represents the extension of the spring.

Question 14

A bag of mass 5.0 kg is lifted from the ground up to a bench that is 1.1 m high. How much gravitational potential energy does the bag gain?

A 54 J

B 55 J

C 5.4 J

D 5.5 J

Question 15

A 300 g ball, thrown straight up with an initial velocity of 11 m s^{-1}, will reach a height of:

A 3.1 m above the ground.

B 6.2 m above the ground.

C 7.1 m above the ground.

D 9.1 m above the ground.

Question 16

Particles in a longitudinal wave:

A move up and down as the wave travels through.

B move backwards and forwards as the wave travels through.

C stay in a fixed position as the wave passes through.

D travel along with the wave.

Question 17

When a light ray travels from a medium of lower refractive index to one of higher refractive index, it:

A bends towards the normal and its speed increases.

B bends towards the normal and its speed decreases.

C bends away from the normal and its speed increases.

D bends away from the normal and its speed decreases.

Question 18

Evidence for the wave-like interaction of light with matter includes:

A refraction, reflection and polarisation.

B refraction and polarisation, but not reflection.

C refraction and reflection but not polarisation.

D polarisation, but not reflection or refraction.

Question 19

What is the critical angle for light travelling from diamond ($n = 2.42$) to crown glass ($n = 1.52$)?

A 37°

B 39°

C 41°

D 43°

9780170412551

Question 20

A ray of light travelling through some crown glass (n = 1.52) enters some flint glass (n = 1.65). If the incident angle is 26°, what is the angle of refraction?

A 21.0°

B 22.8°

C 23.0°

D 23.8°

SHORT-RESPONSE QUESTIONS

Question 1

23°C on the Kelvin scale is:

Question 2

An A380 aircraft uses 15 700L of fuel during a flight at 38% efficiency. How much fuel would the A380 need for the same flight if it were flying at 100% efficiency?

Question 3

Name the process by which two light nuclides combine.

Question 4

What is the SI unit for effective dose of radiation measured in?

Question 5

In the electrical circuit below the reading for V_2 is 7.0 V. Determine the reading for V_1 and state the law used to find this value.

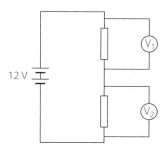

Question 6

Naturally occurring rubidium comprises two isotopes: $^{85}_{37}$Rb, the most common isotope, occurring 72% of the time, and $^{87}_{37}$Rb, occurring 28% of the time. Determine its relative atomic mass.

Question 7

A radioactive isotope has a half-life of 20 days. How long will it take for only one-eighth of the isotope to remain?

Question 8

Describe the term 'nuclear binding energy'.

Question 9

Determine the mass equivalence of $^{104}_{43}$Tc in kilograms. Recall that 1 u = 1.660×10^{-27} kg; mass $_{proton}$ = 1.0078 u; mass $_{neutron}$ = 1.0086 u.

Question 10

A steady current of 2.0 A flows in a wire. How many electrons per second does this represent?

Question 11

State a formula used to determine the displacement of an object.

Question 12

A car takes 5 hours to cover 450 km. Determine its average speed in both $km\,h^{-1}$ and $m\,s^{-1}$.

Question 13

A small hammer of mass 0.75 kg falls from a 4.2 m high roof. How fast will the hammer be travelling when it reaches the ground?

Question 14

The period of a particular electromagnetic wave is 4.0 ms. Determine the frequency of the wave.

Question 15

State the law of reflection.

Question 16

A child steps onto the road 30 m ahead of a car travelling at 60 $km\,h^{-1}$. The driver reacts over a period of 0.2 s and manages to pull up just in time. Find the average acceleration of the car while the brakes were being applied.

9780170412551

Question 17

Weight and mass differ. Describe the difference between mass and weight, including their units of measurement.

_____

Question 18

A 1.5 kg shot-put lands on a large spring with a spring constant of $700 \, N\,m^{-1}$. It has a speed of $3.0 \, m\,s^{-1}$ as it makes contact with the spring. Determine how far the spring will be compressed before the shot momentarily comes to rest.

_____

Question 19

An FM radio station has a base frequency of 102.1 MHz. Determine the wavelength of the electromagnetic wave.

_____

Question 20

Determine the length of an open-ended air column producing a fundamental frequency of 520 Hz. Use the speed of sound in air of $340 \, m\,s^{-1}$.

_____

COMBINATION-RESPONSE QUESTIONS

Question 1

Refer to the values provided to respond.

Specific heat capacity of solid water (ice) $= 2100 \, J\,kg^{-1}\,K^{-1}$

Specific heat capacity of liquid water $= 4200 \, J\,kg^{-1}\,K^{-1}$

Latent heat of fusion $= 334 \, kJ\,kg^{-1}$

How much energy is required to heat a 5.0 kg bag of ice from $-4°C$ to $15°C$?

_____

_____

Question 2

750 g of copper is brought to melting point. Determine how much energy is required to melt all the solid copper to liquid. The latent heat of copper is $205 \, kJ\,kg^{-1}$.

_____

_____

_____

Question 3

The carbon-14 content of an ancient piece of wood was found to be 3.1% of that of living trees. Assuming that, over the ages, the ratio of carbon-14 to carbon-13 has remained the same in the atmosphere, determine the age of the ancient piece of wood. The half-life of carbon-14 is 5730 years.

_____

_____

_____

Question 4

In a thermal nuclear power station, a single neutron causes uranium-235 (mass = 235.044 u) to undergo fission. The fission fragments have masses of 130.896 u and 102.950 u respectively and two neutrons of mass 1.0086 u are released. How much energy, in MeV, is released in this fission reaction? Recall that 1 u = 931.5 MeV.

_____

_____

_____

Question 5

A 240 Volt RMS household lighting circuit has a circuit breaker that is designed to activate when the RMS current exceeds 8.0 A. Determine the maximum number of 75 W light globes that can be safely placed in parallel within the circuit.

_____

_____

_____

Question 6

In an experiment to verify Hooke's law, a force was applied to a spring and the extension measured. The following data were collected.

FORCE APPLIED TO SPRING (N)	SPRING EXTENSION (m)
20	0.30
40	0.62
50	0.76
60	0.92

a Plot the data on a graph to show how the spring obeys Hooke's law.

b Determine the spring constant for the spring, in SI units.

_____

c Calculate the additional energy stored in the spring as elastic potential energy when it is stretched from 40 cm to 80 cm.

_____

Question 7

A cargo ship accelerates from rest at a constant rate for 10 minutes until it reaches a speed of 18 m s^{-1}. It continues to travel in a straight line for a further 25 minutes at 18 m s^{-1}.

a   Calculate the ship's acceleration for the first 10 minutes, in m s^{-2}.

_____

b   Determine the total displacement of the ship over the 35 minute period.

_____

_____

Question 8

A 50 kg person takes a ride down a slippery slide that begins at a height of 15 m above the ground.

a   Calculate the potential energy associated with the person at the top of the slide.

_____

b   Determine their maximum kinetic energy gained when sliding to the bottom of the slide.

_____

c   Calculate the maximum velocity of the rider.

_____

d   At the bottom of the slide, a frictional force of 1000 N is applied. What distance is required for the person to come to rest?

_____

Question 9

Light travelling from a shard of glass to water ($n = 1.33$) is refracted. If the angle of incidence is 45° and the angle of refraction is 38°, determine the refractive index of the glass.

_____

Question 10

A 4.5 cm object is 14 cm from a converging lens of focal length 7.5 cm. Determine the nature, position and magnitude of the image.

_____

ANSWERS

1.1 KINETIC PARTICLE MODEL OF MATTER

1 a Matter is defined as anything with physical substance.

b Brownian motion is the random motion of small objects suspended in a fluid resulting from the objects being bombarded by the particles of the fluid.

c The atom is the fundamental building block of matter.

d A molecule is a collection of atoms bound together by chemical bonds.

e An elastic collision is a collision between two or more objects in which there is no loss of kinetic energy.

f Kinetic energy is the energy of an object due to its motion.

g Potential energy is the energy that is stored in a system due to the configuration and interaction of the bodies with the system.

2

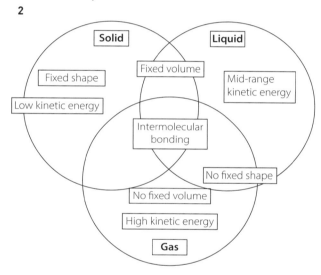

3 The relationship shown is clearly linear as is supported by the equation: $E_{k\,av}T$. That said, since only two points are present, no definite decision can be made on the relationship. The gradient of the line represents the constant of proportionality between $E_{k\,av}$ and T. The equation could read: $E_{av} = 6.9 \times 10^{20} \times T$.

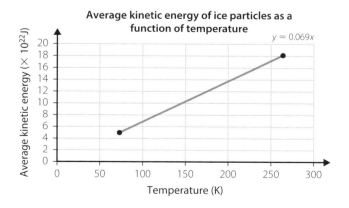

1.2 THE ENERGY MODEL

1 a All forms of energy can be *transformed* from one form to another and *transferred* from one place to another.

b In an *elastic* collision, the total energy of the colliding particles is always *conserved*.

c *Kinetic* energy is the energy a body possesses due to its motion.

d The potential energy of a substance is the energy that is stored in the way that the particles are connected to each other through the existence of *intermolecular bonds*.

e The *internal* energy of a substance is equal to the *sum* of the kinetic and potential energies of its particles.

f The *temperature* of a substance is a measure of the average kinetic energy of its particles.

g *Heat* is energy that spontaneously moves between substances because of a difference in temperature between them.

2 Sections 1, 3 and 5 refer to sections where the temperature of the substance is increasing while being heated. This means that the average kinetic energy of the substance is increasing and the total internal energy is also increasing.

Sections 2 and 4 refer to sections where the temperature of the substance is remaining constant while heat energy is being added. This means that the potential energy stored in the intermolecular bonds must be increasing, and therefore, so is the internal energy of the substance.

1.3 HEAT TRANSFERS

1

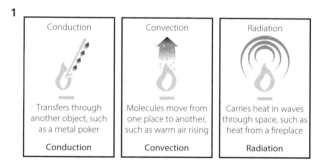

Conduction	Convection	Radiation
Transfers through another object, such as a metal poker	Molecules move from one place to another, such as warm air rising	Carries heat in waves through space, such as heat from a fireplace
Conduction	Convection	Radiation

2

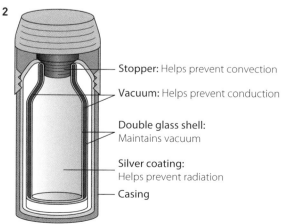

Stopper: Helps prevent convection

Vacuum: Helps prevent conduction

Double glass shell: Maintains vacuum

Silver coating: Helps prevent radiation

Casing

3 **a** Conduction   **b** Convection

     c Conduction   **d** Radiation

     e Radiation and convection   **f** Convection

CHAPTER 1 EVALUATION

1 D   **2** A   **3** B

4 D   **5** Heat   **6** Temperature

7 The kinetic and potential energies of individual particles within a substance are constantly undergoing elastic collisions with each other. During these collisions, energy is being transferred between particles. As the energy is transferred from one part of the substance to another, the total amount of energy in the substance (the internal energy) will remain constant.

8 The substance must be undergoing a phase change and all of the added heat is being stored as potential energy by moving the particles away from their ideal intermolecular bond length. If the heat was being transformed to kinetic energy, the temperature would rise.

9 Metals are good conductors since they will quickly transfer heat across them. This results in them heating up very quickly and conducting the increased kinetic energy to another area very quickly. This property is related to the large number of delocalised electrons that are free to gain kinetic energy.

10 The heat from the Sun undergoes radiation in the EM spectrum across the vacuum between the Sun and the Earth. When the heat enters the atmosphere, the air particles will undergo convection and conduction. Some heat will reach a body of water, whose particles undergo conduction and convection until they have sufficient energy to evaporate. Once the water particles enter the atmosphere, convection will take the warmed water particles into the sky, where they will undergo conduction with neighbouring air particles. This process results in them losing energy and condensing out of the gaseous phase as rain drops.

CHAPTER 2 REVISION

■ 2.1 CONVERTING TEMPERATURE

1

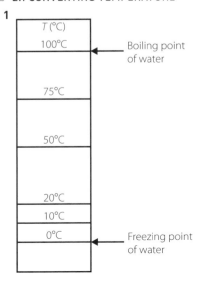

2 Lord Kelvin *hypothesised* that the volume of an *ideal* gas should condense to zero at −273°*Celsius*. The only way that this could occur is if the *kinetic* energy of the particles reduced to zero. He used this *absolute* temperature as the basis for his new temperature scale, the *Kelvin* scale. Unfortunately, the *second* law of thermodynamics states that heat always moves from a *hotter* object to a *colder* object, which means that we can never reach this *lowest* of all temperatures, because, to cool down, the object would need to *transfer* its energy to something colder than it. So to reach absolute zero, we would need another *object* that is already at absolute zero, and we are yet to find such an object.

3

PROPERTY	T_C (°C)	T_K (K)
The freezing point of water	0	273
The boiling point of water	100	373
The freezing point of milk	−4	269
The average temperature of space	−270.3	2.7
Average surface temperature of Earth	15	288
Average surface temperature of the Sun	5505	5778
Average core temperature of the Sun	1.57×10^7	1.57×10^7

■ 2.2 COLLECTING DATA

TYPE OF THERMOMETER	DESCRIPTION
Thermostat	Uses variation in electrical resistivity of a material with temperature
Mercury in glass	Uses different coefficients of expansion between mercury and glass
Bimetallic strip	Uses variation in coefficients of expansion between two different metals to detect temperature changes
Thermal paint	Uses colour change with temperature
Digital	Uses the variation in resistivity of a material with temperature; the greater the resistance the lower the current
Thermocouple	Uses different temperature-dependent electrical properties of different metals that are brought into contact
Infrared	Uses the electromagnetic radiation from a surface to measure temperature on the absolute temperature scale

■ 2.3 PRACTICAL SKILLS: MEASUREMENTS

1

TERM	DEFINITION
Measurand	A specified quantity to be measured
Measurement result	The best estimate of the true value given the limitations of the actual measurement device used
True value	For continuous variables, this is an unknowable ideal value that represents the measurand
Precision	The degree to which the individual measurements cluster around their mean value
Accuracy	The degree to which a measurement correctly reflects or approaches the true value
Uncertainty	The estimate of the range of values within which the true value of a measurement or derived quantity lies
Indication values	Single results of a measurement
Mean value	The average value of a set of indication values
Accepted value	The value of a substance or quantity that is universally agreed as being a best estimate due to multiple and highly accurate measurements
Scientific form	A means of representing numbers as products of exponents
Random error	A variation that affects a measurement in a random way so that successive measured values may reflect small changes from each other
Systematic error	An error that acts in a predictable manner to give a consistent offset in data
Absolute uncertainty	The size of the range of values in which the true value of a measurement probably lies
Parallax error	Error in a measurement caused by the change in the apparent position of an object viewed from two different lines of sight
Confidence interval	A range of values in which an indication value lies
Maximum value	Upper limit of a confidence interval
Minimum value	Lower limit of a confidence interval
Significant figures	Digits reported in a measurement result; the number of significant figure is the number of meaningful digits in a measurement result
Percentage uncertainty	A measure of the uncertainty of a measurement compared to the size of the measurement, given as a percentage
Relative uncertainty	A measure of the uncertainty of a measurement as a fraction of the measurement result
Proportional error	The difference between a measurement result and an accepted value, expressed as a fraction of the accepted value
Percentage error	The difference between a measurement result and an accepted value, given as a percentage of the accepted value
Fundamental units	The seven units defined for the fundamental or basic quantities of length, mass, time, current, temperature, luminous intensity and amount of matter
Derived units	Units derived from a combination of the fundamental units

2 a i $0.45 \pm 0.005°C$

   ii $99.35 \pm 0.005°C$

 b i 1.1%

   ii $5.0 \times 10^{-3}\%$

 c i 0.65%

■ 2.4 CHANGES IN TEMPERATURE

1

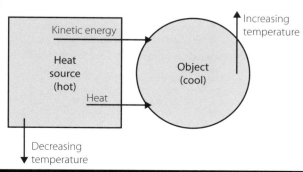

2

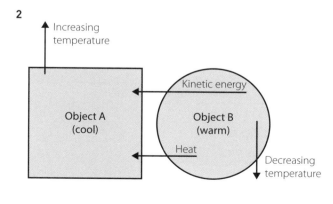

2.5 SPECIFIC HEAT CAPACITY AND PROPORTIONALITY

1

SUBSTANCE	SPECIFIC HEAT CAPACITY $(J kg^{-1} K^{-1})$	ORDER (1–6)
Steam	2000	4
Cooking oil	2800	2
Soil	800	6
Ice	2100	3
Liquid water	4180	1
Air	1000	5

2 a

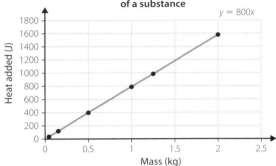

Heat required to increase the temperature of a substance

$y = 800x$

b The linear relationship shows that there is a direct proportionality between mass and heat required to increase the temperature of the substance by 1°C.

c $800 J kg^{-1} K^{-1}$

d Soil

2.6 SOLVING PROBLEMS: SPECIFIC HEAT CAPACITY

1 $2820 J kg^{-1} K^{-1}$

2 $25000 J$

3 $1.92 \times 10^3 J$

4 $6.6 kg$

5 $620 J kg^{-1} K^{-1}$

2.7 INTERPRETING SPECIFIC HEAT DATA

1

MEASUREMENT	VALUE
Mass of calorimeter (g)	150
Mass of cold water and calorimeter (g)	312
Mass of water (g)	162
Voltage (V)	11.5
Current (A)	2.1
Power (W)	24.15

2 a

TIME (s)	HEAT ADDED (J)	TEMPERATURE (°C)
0.0	0	15.00
30.0	725	16.00
60.0	1449	16.90
120.0	2898	18.85
150.0	3623	19.80
180.0	4347	21.10
228.0	5506	22.35
267.0	6448	23.6
306.0	7390	24.85
345.0	8332	26.10
384.0	9274	27.35
423.0	10215	28.60
462.0	11157	29.90
501.0	12099	31.15
540.0	13041	32.40
579.0	13983	33.65
618.0	14925	34.90
657.0	15867	36.15

b–d

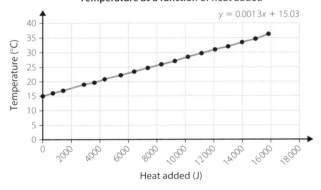

Temperature as a function of heat added

$y = 0.0013x + 15.03$

e $4748 J kg^{-1} K^{-1}$

f 13.6%

CHAPTER 2 EVALUATION

1 C **2** C **3** C **4** A **5** A

6 Qualitative

7 Systematic

8 Half as much

9 A common scale is useful for communicating ideas and the Kelvin scale does not include negative numbers.

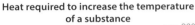

10 Most substances occupy a larger volume at higher temperatures; therefore, if there were no expansion gaps, there is a chance that the concrete would crack during the heating and cooling processes.

11 $0.091 \, \text{kcal} \, \text{kg}^{-1} \, \text{K}^{-1}$

12 $23\,000 \, \text{J}$

13 $740 \, \text{J} \, \text{kg}^{-1} \, \text{K}^{-1}$

14 $27°\text{C}$

15 $16.9°\text{C}$

16 Student response should contain reference to an increase in kinetic energy at a surface on application of heat and the transfer of this kinetic energy to other particles within the substance through the process of elastic collision. Reference should also be made to an increase in the average kinetic energy of the substance.

CHAPTER 3 REVISION

■ 3.1 THE PROCESS OF STATE CHANGE

1

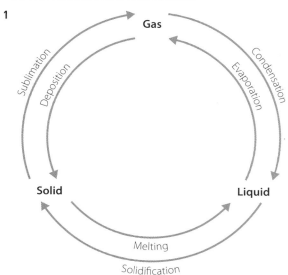

2

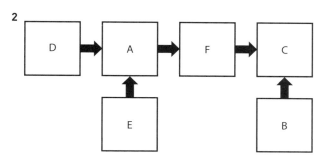

3

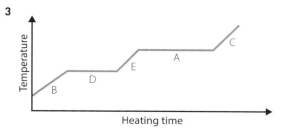

■ 3.2 DEFINING SPECIFIC LATENT HEAT

1 a

MASS OF ICE AT 0°C (kg)	TIME FOR MELTING (s)	TOTAL ENERGY INPUT (kJ)
0.10	66	33
0.22	144	72
0.39	176	88
0.52	350	175
0.64	434	217
0.90	600	300

b

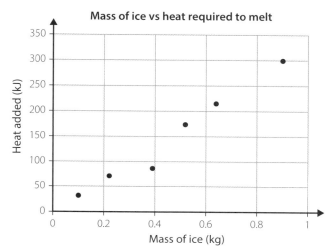

c The heat required to melt the 0.39 g sample is an anomaly since the remaining data points show a linear relationship. This could be caused by an error in a reaction time, impurities in the sample, inaccurate determination of the point when the sample had completely melted, or due to an inaccurate determination of the mass of the sample.

d

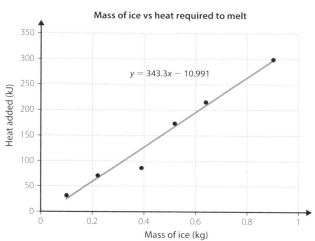

e $y = 336.08x - 0.5741$ or $Q\,(\text{kJ}) = 336.08x\,\text{m\,(kg)}$

f The gradient would refer to the latent heat of fusion for ice.

g The accepted value of L_{fusion} for ice is $334\,\text{kJ\,kg}^{-1}$; this is in very close agreement with the experimentally derived value of $343\,\text{kJ\,kg}^{-1}$.

■ 3.3 SOLVING PROBLEMS: SPECIFIC LATENT HEAT

1 $9.0 \times 10^2\,\text{kJ}$

2 $96.06\,\text{g}$

3 $13\,\text{kJ}$

■ 3.4 PHASE CHANGES

1 a

TIME (s)	TEMPERATURE (°C)	HEAT ADDED (kJ)
0	−10	0
60	−1.7	3.6
120	0	7.2
180	0	10.8
240	0	14.4
300	0	18
360	0	21.6
420	0	25.2
480	0	28.8
540	0	32.4
600	0	36
660	0	39.6
720	0	43.2
780	0	46.8
840	0	50.4
900	0	54
960	0	57.6
1020	0	61.2
1080	0	64.8
1140	0	68.4
1200	1	72
1260	5.3	75.6
1320	9.6	79.2
1380	10.1	82.8
1440	18.1	86.4

b

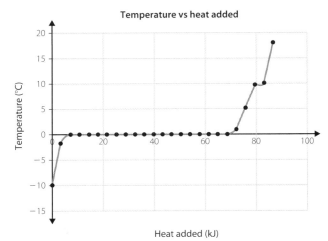

Temperature vs heat added

c An anomalous point exists at 1380 s where the temperature remains almost constant despite heat being added after the completion of the phase change. This may be due to incomplete melting of the sample.

d Taking the melting phase from 120 s to 1140 s, the heat added during this time is 61.2 kJ. Since the mass of the sample was 150 g, the latent heat of fusion is $408\,\text{kJ\,kg}^{-1}$. The accepted latent heat of fusion for water is $334\,\text{kJ\,kg}^{-1}$.

e The significant difference in the accepted latent heat of fusion and the calculated latent heat of fusion could be due to insufficient control of variables, such as the thermodynamic isolation of the sample from the surroundings. In this case, heat would have escaped to the environment. In the future, it is suggested that a higher quality calorimeter be used. In addition, the time steps of 1 minute are insufficient to exactly determine the beginning and end points of the phase change. It is suggested that in future tests, temperature readings be taken at smaller intervals.

CHAPTER 3 EVALUATION

1 C   **2** B   **3** D

4 Deposition

5 Specific latent heat of fusion

6 In evaporation, individual particles of a liquid become sufficiently energetic to leave the liquid phase when it is below the boiling point because of random collisions. This results in a lowering of the average kinetic energy of the substance. In vaporisation, because of added heat, all of the particles become sufficiently energetic to overcome the intermolecular bonding present in the liquid phase, and the substance will vaporise.

9780170412551

7

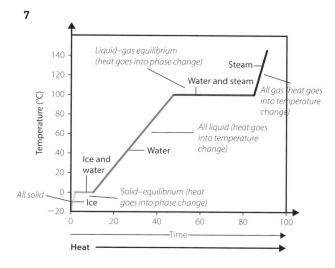

8 470 kJ

9 38 000 J

10 96 kJ kg^{-1}

11 Heat released by steam = 452 kJ; heat required by aluminium = 390 kJ. So, yes – there is sufficient heat.

CHAPTER 4 REVISION

■ 4.1 THERMAL EQUILIBRIUM AND THE ENERGY OF PARTICLES

1 Two objects that are at the same temperature are said to be in *thermal* equilibrium.

A hotter object will transfer *kinetic energy* to a cooler object through elastic *collisions*.

The *heat* lost by the hotter object is equal to the heat gained by the cooler object.

At thermal equilibrium, the net heat flow between the objects is equal to *zero*, and the average kinetic energy of the two objects is *equal*.

2

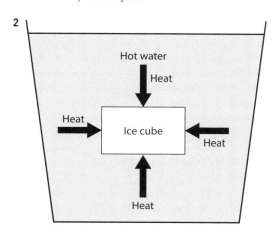

Heat flows from the hot water into the ice cube, which results in a decrease in the average kinetic energy and temperature of the water, and an increase in the average kinetic energy and temperature in the ice cube.

3

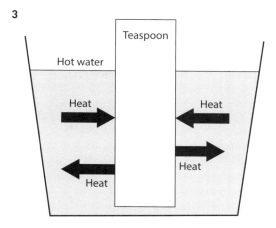

Heat flows both from the hot water into the teaspoon and from the teaspoon into the hot water. The net heat flow in one direction is equal to the net heat flow in the other, meaning that the average kinetic energy and, therefore temperature, remains constant in both.

■ 4.2 ACHIEVING THERMAL EQUILIBRIUM

1

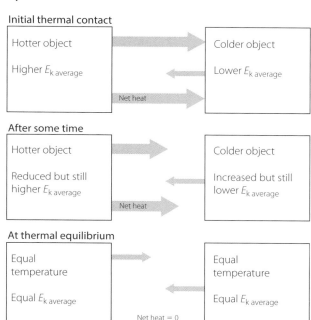

2 The *zeroth law of thermodynamics* states that if two objects are in *thermal equilibrium* with a third *object*, then they must be in thermal equilibrium with each other. The importance of the zeroth law is that it gives a useful description of *temperature* that agrees with our *everyday experience* that, when a hot and a cold object are put into *thermal contact*, they will eventually reach the same temperature.

■ 4.3 SOLVING PROBLEMS: THERMAL EQUILIBRIUM AND THE SPONTANEOUS TRANSFER OF HEAT

1

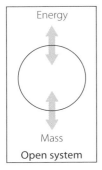
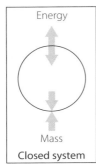
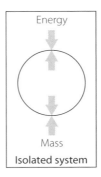

2 **a** Isolated

   b Closed

   c Open

   d Closed

   e Isolated

   f Open

3 32°C

4 49 g

5 $120\,\mathrm{J\,kg^{-1}\,K^{-1}}$

CHAPTER 4 EVALUATION

1 B

2 A

3 C

4 Temperature and average kinetic energy

5 Increase

6 Open system

7 Temperature

8 Student responses may vary, but should make reference to transfer of heat through conduction, elastic collisions, and equilibration of average kinetic energy.

9 Student responses may vary but should make reference to the fact that there are still collisions occurring between the two substances.

10 If the two objects are of different substances, the two substances might have different specific heat capacities and would therefore have different temperature changes associated with an equal input of heat.

11 50°C

12 55°C

13 94 g

14 $2700\,\mathrm{J\,kg^{-1}\,K^{-1}}$

CHAPTER 5 REVISION

■ 5.1 THE CAPACITY TO DO WORK

1 In a *closed* system, *energy* can be added or removed by either *adding* or removing *heat* or by the system doing *work* or having work done on it. An external *combustion* engine is an example of such a closed system, whereby *external heat* is added to the system by the *burning* of a fuel in an external combustion *chamber*, which then heats a *fluid* that expands and does *work* to move a *piston*. This motion can be used to do work on something that is *external* to the system.

2

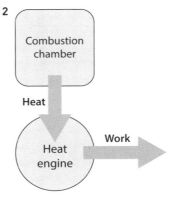

■ 5.2 CHANGE IN INTERNAL ENERGY

1

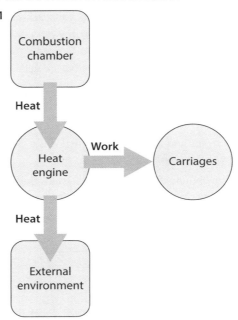

2 3600 J

3 $6.0 \times 10^5\,\mathrm{J}$

■ 5.3 HEAT LOSS AND USABLE ENERGY

1 The gas inside the evaporator coil of an air conditioner unit is kept cooler than the outside air temperature, meaning that heat can be transferred from outside into the evaporator coil. This gas is then passed through a compressor, which transforms the gas into a warm liquid that is warmer than the internal air temperature. Heat will move out of the liquid into the internal atmosphere. The liquid is then passed through a condenser that returns it to a cool gas.

9780170412551

2

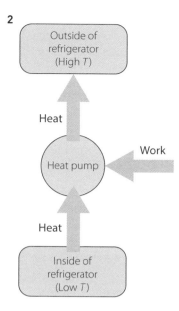

3

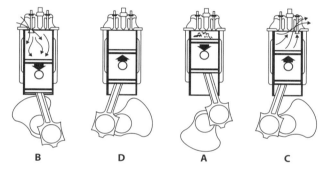

B   D   A   C

5.4 EFFICIENCY

1 48%

2 6900 J

3 a 28 kJ

b 5500 kJ

1 C

2 A

3 D

4 A

5 B

6 Internal

7 Increasing

8 $P = \dfrac{W}{t} = \dfrac{Fd}{t} = F\dfrac{d}{t} = Fv$

9 Heat moves into the engine by the combustion of fuel, and this results in work being done on the piston.

10 Heat moves into the system from the internal atmosphere; work is done to the system to compress the refrigerant; heat moves out of the system to the external atmosphere.

11 26 000 J

12 330 kJ

13 92%

14 12 kJ

6.1 ISOTOPES

1 An element is defined by the number of protons in its nucleus. An isotope is a subcategory of an element, and is further defined by the number of neutrons in the nucleus.

2 Protium ($A = 1$), deuterium ($A = 2$) and tritium ($A = 3$)

3 $^{1}_{1}\text{H}$, $^{2}_{1}\text{H}$, $^{3}_{1}\text{H}$

4 Atomic weight is found by adding the percentage abundance multiplied by the mass number of each isotope:

$= 0.93 \times 39 + 0.07 \times 41$

$= 39.14$

5 Equate the atomic weight to the formula for weighted average and solve:

$35.45 = 0.76 \times 35 + 0.24x$

$0.24x = 8.85$

$x = 36.88$

$x \approx 37$

The other isotope of chlorine in this sample is chlorine-37.

6.2 NUCLIDES

1 The number of nucleons is read as the top number, number of protons is read as the bottom number, and the number of neutrons is the difference between the number of protons and neutrons.

a 14 nucleons, 8 neutrons, 6 protons

b 41 nucleons, 22 neutrons, 19 protons

c 7 nucleons, 4 neutrons, 3 protons

d 63 nucleons, 34 neutrons, 29 protons

e 24 nucleons, 12 neutrons, 12 protons

f 235 nucleons, 143 neutrons, 92 protons

g 78 nucleons, 42 neutrons, 36 protons

2 The number of nucleons can be determined as the sum of the proton and neutron numbers, if it is not given. The proton number is the difference between the nucleon and neutron numbers, if it is not given.

a $^{210}_{85}\text{At}$

b $^{188}_{76}\text{Os}$

c $^{45}_{20}\text{Ca}$

d $^{70}_{32}\text{Ge}$

e $^{144}_{60}\text{Nd}$

3 It would quadruple.

6.3 THE STABILITY CURVE

1 $^{20}_{10}\text{Ne}$, $\dfrac{\text{n}}{\text{p}} = \dfrac{10}{10} = 1$

$^{21}_{10}\text{Ne}$, $\dfrac{\text{n}}{\text{p}} = \dfrac{11}{10} = 1.1$

$^{22}_{10}\text{Ne}$, $\dfrac{\text{n}}{\text{p}} = \dfrac{12}{10} = 1.2$

2 Check with your teacher. Solution should include that more protons will increase the electrostatic force, and hence more neutrons would be required not only to increase the distance between the protons, but also to increase the strong nuclear force within the nucleus.

3 a Yes

b No

c No

d Yes

1 C

A is incorrect. B and D are true; however, they are not the sole reason the nucleus stays together.

2 D

3 Atomic weight is found from the addition of the percentage abundance multiplied by the mass number of each isotope:

$= 0.95 \times 32 + 0.045 \times 34 + 0.0075 \times 33$

$= 30.4 + 1.53 + 0.2475$

$= 32.18$

The relative atomic mass of sulfur is 32.18.

4 When the protons are distance d apart, they exert the following force on each other:

$$F_1 = \frac{kq_1q_2}{d_1^2}$$

When the protons are moved closer together, the force they exert on each other can be represented as:

$$F_2 = \frac{kq_1q_2}{d_2^2}$$

where $d_2 = \frac{d_1}{3}$

Substituting this in we obtain:

$$F_2 = \frac{kq_1q_2}{\left(\frac{d_1}{3}\right)^2}$$

$$F_2 = \frac{kq_1q_2}{\frac{1}{9}d_1^2}$$

$$F_2 = 9 \times \frac{kq_1q_2}{d_1^2}$$

$$F_2 = 9 \times F_1$$

The force of repulsion between the protons increases by a factor of nine when the distance between them is a third of the original separation distance.

5 Over very short distances the four fundamental forces act differently compared to over very large distances. There is a very small gravitational attraction between protons in the nucleus of an atom since they have mass. There is also an extremely large electrostatic repulsion between these protons since they are in such close proximity. If these were

the only two calculable forces, then nuclei would not be able to stay together. The nuclei of atoms do stay together; however, due to a third force—a gluing force—known as the strong nuclear force, the strong nuclear force also acts between these protons as a force of attraction. This strong nuclear force keeps the nucleus of an atom together due to the meson exchange that is constantly occurring between nucleons. This only happens at very small separation distances, and when the distance is this small, the strong nuclear force is the largest fundamental force.

■ 7.1 TYPES OF RADIATION

1 Alpha particle: 3.2×10^{-19} C, positron: 1.6×10^{-19} C, electron: -1.6×10^{-19} C

2 In order of most penetrating: gamma, beta, alpha

In order of most ionising: alpha, beta, gamma

3 Some nuclides do not have a stable number neutron to proton ratio, so they eject particles from the nucleus in an attempt to obtain this.

4 Too many protons in the nucleus will cause positron decay; too many neutrons in the nucleus will cause beta-minus decay, and if there are too many nucleons the nucleus will eject an alpha particle.

5 a $^{241}_{95}\text{Am} \rightarrow \, ^{237}_{93}\text{Np} + \, ^{4}_{2}\alpha$

b $^{90}_{38}\text{Sr} \rightarrow \, ^{90}_{39}\text{Y} + \, ^{0}_{-1}\beta$

c $^{11}_{6}\text{C} \rightarrow \, ^{11}_{5}\text{C} + \, ^{0}_{1}\beta$

d $\dfrac{1}{R_T} = \dfrac{1}{R_1} + \dfrac{1}{R_2} + \ldots + \dfrac{1}{R_n}$

6 a Lead-210

b Bismuth-210

c Polonium-216

d Nitrogen-15

7 a Beta-minus decay

b Beta-positive decay

c Beta-minus decay

d Alpha decay

■ 7.2 HALF-LIFE

1 12.5% means the $\dfrac{N}{N_0}$ ratio is $\dfrac{1}{8}$.

$$\frac{N}{N_0} = \left(\frac{1}{2}\right)^n$$

$$\frac{1}{8} = \left(\frac{1}{2}\right)^n$$

$$n = \log_{\frac{1}{2}}(1/8)$$

$n = 3$ half-lives

If each half-life is 20 minutes, it means that it will take 60 minutes for only 12.5% of the original nuclei of this sample to remain.

9780170412551

2 $\dfrac{N}{N_0} = \left(\dfrac{1}{2}\right)^n$

$\dfrac{1}{16} = \left(\dfrac{1}{2}\right)^n$

$n = \log_{1/2}(1/16)$

$n = 4$ half-lives

3 $N = N_0\left(\dfrac{1}{2}\right)^n$

$N = 150\left(\dfrac{1}{2}\right)^7$

$N = 1.17\text{g}$ remaining

4 If $N = N_0\left(\dfrac{1}{2}\right)^n$ and N and A are directly proportional, then $A = A_0\left(\dfrac{1}{2}\right)^n$

First, work out the number of half-lives:

$\dfrac{A}{A_0} = \left(\dfrac{1}{2}\right)^n$

$\dfrac{10}{100} = \left(\dfrac{1}{2}\right)^n$

$n = \log_{1/2}(10/100)$

$n = 3.32$ half-lives

For time taken:

$3.32 \times 5700 = 18\,934.9$ years

So it would take approximately 19000 years for the activity to fall to 10 counts per minute.

CHAPTER 7 EVALUATION

1 B

2 B

3 B

4 a $^{178}_{73}\text{Ta} \rightarrow\ ^{178}_{74}\text{W} +\ ^{0}_{-1}\beta$

$^{183}_{73}\text{Ta} \rightarrow\ ^{183}_{74}\text{W} +\ ^{0}_{-1}\beta$

Only tungsten-183 is stable; tungsten 178 is not.

b $^{178}_{73}\text{Ta} \rightarrow\ ^{178}_{72}\text{Hf} +\ ^{0}_{1}\beta$

$^{183}_{73}\text{Ta} \rightarrow\ ^{183}_{72}\text{Hf} +\ ^{0}_{1}\beta$

Only hafnium-178 is stable; hafnium-183 is not.

5 $^{72}_{30}\text{Zn} \rightarrow\ ^{72}_{31}\text{Ga} +\ ^{0}_{-1}\beta$

$^{72}_{31}\text{Ga} \rightarrow\ ^{72}_{32}\text{Ge} +\ ^{0}_{-1}\beta$

6

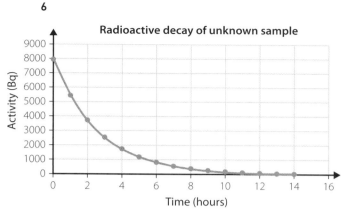

Radioactive decay of unknown sample

Students can choose any sets of data as long as of the two y values chosen, one is half the other. The half-life of this sample is approximately 1.8 hours.

7 Response should include that nuclides are nuclei that are unstable when in their ground state. It is possible for isotopes to have an excited nucleus, but when the nucleus is in the ground state it is stable. Hence, it is more accurate to talk about a nuclide than an isotope, since it considers the energy state of the nucleus, not just its components.

CHAPTER 8 REVISION

■ 8.1 NUCLEAR ENERGY AND MASS DEFECT KEY TERMS

1 A *fission chain reaction* occurs when a *slow (thermal) neutron* is absorbed into the nucleus of a larger element, such as uranium. The now unstable element splits into several *fission fragments*, with a total mass that is less than the mass of the element initially. This difference in mass is termed the *mass defect*, and is able to be used to calculate the energy released in the interaction, using *Einstein's mass–energy equivalence equation*. The resulting release of energy may be used to generate electricity through a *controlled nuclear chain reaction*.

2 Student answers will vary but should include the following ideas.

- The four fundamental forces act at different scales. The weak nuclear force and the strong nuclear force act within the nucleus of the atom, while the electromagnetic force acts across both atomic and macroscopic scales. The force of gravitation is quite weak, when compared to the other forces, and is most significant on a larger, planetary scale.

- Nuclear fission and fusion are processes that occur within the nucleus of atoms. In nuclear fission, a heavy nucleus is split into new fission fragments or elements with differing numbers of nucleons and an accompanying release of energy. In nuclear fusion, lighter nuclei are fused together to make a new element, again with an accompanying release of energy.

8.2 ENERGY SOURCE EVALUATION

TABLE 8.2.1 Energy source evaluation

CRITERIA	NUCLEAR POWER	FUEL SOURCE 2: COAL-BURNING POWER	FUEL SOURCE 3: HYDROELECTRIC POWER
Source availability	The most common nuclear fuel, uranium, is found throughout the Earth's crust and is about 500 times more abundant than gold. It is found in concentrations of about four parts per million (ppm) in granite; however, it is not renewable.	The consumption of coal has been rising every year. At current rates, it is estimated that the coal reserves will only last for one more century. Coal is a non-renewable source of energy. Current reserves of coal were formed 350 million years ago.	Hydroelectricity is a renewable energy source; however, it is geographically dependent upon typical rain and snowfall.
Running cost (per MWh)	Approximately $790/MWh	Approximately $8/MWh	Approximately $110/MWh
Initial cost	The nuclear industry generally reports the cost of construction to be about $2000 per installed kilowatt (kW). However, construction expenses vary widely and depend on the cost of local labour and government regulations, with estimates as low as $1200 per kW and as high as $8000 per kW. The cost of construction to build a nuclear reactor is between US$10.4 and $11.9 billion.	Australia already has the appropriate infrastructure required since coal is currently Australia's dominant source of electricity. The estimated cost of building new coal plants have reached $3500 per kW. This represents $2 billion for a 600 MW coal plant, when financing costs are included.	Australia has several hydroelectric schemes in operation. The Snowy Mountains hydroelectric scheme consists of: sixteen major dams; seven power stations; two pumping stations; and 225 kilometres of tunnels and pipelines. Constructed between 1949 and 1974, the scheme had a cost of A$820 million in 1974 (equivalent to A$6 billion nowadays).
Reliability and suitability	High reliability for base load power.	High reliability for base load power.	High reliability for base load power.
Greenhouse gas emissions	Although nuclear power plants do not emit CO_2, SO_2 or NO in the production of electricity, such emissions are associated with the mining and transport of the uranium itself.	From mining to coal cleaning, from transportation to electricity generation to disposal, coal releases numerous toxic pollutants into the air, water and land. These disrupt ecosystems and endanger human health.	Hydroelectric power plants do not emit CO_2, SO_2 or NO in the production of electricity. However, such emissions are associated with the initial building of the hydroelectric power stations.
Safety	The less obvious or delayed effects of producing nuclear power are known today, and include exposure to cancer-inducing substances and radiation. Health risks from uranium mining are very low. The risks of exposure to radiation are well known, and radioactivity is easily measureable. Long-term risks are well understood, and can be minimised. Natural sources account for most of the radiation we receive.	There are immediate risks for workers, since coal must be mined and transported. A large amount of coal is required to supply a power plant—approximately 20 000 times more coal than uranium must be mined. Mining and multiple handling of so much material involves hazards. There are approximately 200 times the number of immediate fatalities resulting from producing electricity from coal than from nuclear power. Coal plants produce vast amounts of air pollutants, including carbon dioxide, sulfur dioxide, carbon monoxide, mercury, arsenic and lead.	There are no immediate risks for workers or the community when using hydroelectric power.

9780170412551

1 a i 57.6 N

   ii 14.4 N

b 57.6 N (equivalent to the electrostatic repulsion)

2 Additional neutrons within the nucleus provide greater distance between protons and hence decrease the electrostatic force of repulsion between them.

3

TABLE 8.3.1 Stability chart of light elements

NUMBER OF NEUTRONS, N	1	2	3	4	5	6	7	8
16						C-16		O-16
15						C-15		
14						C-14		
13						C-13		
12						C-12		
11						C-11		
10						C-10		
9						C-9		
8								
7								
6								
5								
4								
3								
2		He-3						
1								
0	H-1							

ATOMIC NUMBER, Z

4 The amu is defined as $\frac{1}{12}$ of the mass of the C-12 isotope. Hence, 1 amu is equivalent to $\frac{1}{12} \times$ mass C-12 or 1.66×10^{-27} kg. Using $E = mc^2$, the energy equivalent of this mass is 1.49×10^{-10} J, or 931.3 MeV.

5

PARTICLE	MASS		
	kg	u	MeV/c^2
Proton	1.673×10^{-27}	1.007 28	941.1
Neutron	1.674×10^{-27}	1.008 67	941.6
Electron	9.11×10^{-31}	0.000 55	0.511

8.4 THE FOUR FUNDAMENTAL FORCES

1

TABLE 8.4.1 The four fundamental forces

	TYPE OF FORCE			
	GRAVITATIONAL	WEAK NUCLEAR	ELECTROMAGNETIC	STRONG NUCLEAR
Relative magnitude	1	10^{32}	10^{36}	10^{40}
Range (m)	Infinite	10^{-18} or 1 attometre, 1 am	Infinite	10^{-15} or 1 femtometre, 1 fm

2 a *Protons* have a positive charge.

b *Like* charges repel, whereas *unlike/opposite* charges attract.

c The repulsion that exists between protons within the nucleus is the result of *electrostatic* force.

d The force that keeps nucleons together is termed the *strong nuclear force*.

e Einstein's mass–energy equation is quantitatively stated as $E = mc^2$.

3 a 5.04×10^{-29} kg

b 7.085 MeV

4 a $^2_1H + \ ^2_1H \rightarrow \ ^3_2He + \ ^1_0n + energy$

b $^2_1H + \ ^3_1H \rightarrow \ ^4_2He + \ ^1_0n + energy$

5 A slow thermal neutron has less kinetic energy than a fast neutron, following collisions with a moderator. Slow neutrons are more effectively captured within the nucleus of atoms, allowing a sustained nuclear chain reaction to be achieved.

6 A slow neutron is captured by a U-235 molecule, which then undergoes fission, releasing three neutrons, some of which continue and take part in further nuclear fission reactions.

4 Low-level waste: 1949 m^3

Intermediate-level waste: 447 m^3

High-level waste: nil

5 There are approximately 100 locations.

6 A single storage site would allow safe and efficient storage and would require approximately 40 hectares to store 100 years' worth of low-level waste.

7 Radioactive decay continues for all radioactive products at different rates. The half-lives of radioisotopes differ due to varied activities. In a period of one half-life, the amount of radioisotope decreases to half the original amount.

8

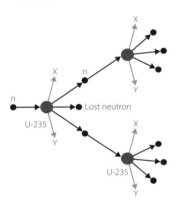

TABLE 8.5.1 Nuclear radiation characteristics

RADIATION TYPE	PHYSICAL ENTITY	CHARGE	RELATIVE PENETRATION ABILITY	ENVIRONMENTAL IONISATION
Alpha	Helium nucleus	2+	Low	Highly ionising
Beta	High-speed electron	1−	Medium	Ionising
Gamma	Electromagnetic radiation	Nil	High	Nil

8.5 RADIOACTIVE WASTE MANAGEMENT

1 • Low-level waste: Paper, plastic, gloves—items with small amounts of radioactivity

• Intermediate-level waste: Radioactive waste with a contact dose rate of 2 millisieverts per hour and above, from radiopharmaceutical production and reactor operation.

• High-level waste: Waste from the operation of nuclear power plants.

2 a International Atomic Energy Agency (IAEA).

b Australian Radiation Protection and Nuclear Safety Agency (ARPANSA).

3 The sievert, Sv

9 Geological stability is important since high-level waste is radioactive for extended periods of time (thousands of years) and generates a significant amount of heat energy. Disposal in deep formations hundreds of metres below the surface is recognised as most appropriate.

10 Australia does have significant areas of great geological stability and of relatively low population density, thereby offering storage solutions for domestic as well as international nuclear waste. Such a global storage solution does have significant consequences, given the danger and longevity of radioactive waste.

9780170412551

■ 8.6 SAFETY IN NUCLEAR REACTORS

1 Safety-related functions include the systems, structures, components, procedures and controls required to remain functional as a nuclear reactor is shut down.

2

TABLE 8.6.1 Nuclear plant incidents

YEAR	NUCLEAR PLANT	DETAILS OF THE EVENT	APPROXIMATE NUMBER OF FATALITIES
1979	Three Mile Island, United States	Partial meltdown in a reactor following a cooling malfunction. Considered the United States' worst nuclear accident.	Nil
1986	Chernobyl, Ukraine	Widespread health and environmental effects. External release of a significant fraction of reactor core inventory. The explosion of one of the four reactors at the power plant resulted in a fire that burnt for nine days. Significant release of radiation into the air, with contamination reaching nearly every country in the northern hemisphere.	2 people (explosion) 28 people (acute radiation) Thousands of additional cancer-related deaths.
2011	Fukushima, Japan	The Sendai earthquake and resulting tsunami led to the failure of the emergency cooling system at the Fukushima Daiichi nuclear power plant.	2 people (160 000 evacuees in temporary housing since the event.)

3

MINE DISASTER LOCATIONS	NUMBER OF DEATHS
Pike River, New Zealand	29 deaths in 2010
United States of America	1489 deaths in 1900, decreasing to 9 deaths in 2016. 100 000 + in total over the 20th Century
United Kingdom	164 346 in total over the history of coal mining in the United Kingdom
China	1549 deaths in 1942, decreasing to 1049 deaths in 2013 and 931 deaths in 2014

4 **a** A reactor protection system is a set of components in a nuclear reactor that shut down the reactor once activated, either manually or automatically. It functions by introducing all controls rods into the reactor core to shut down the reactor.

b Control rods absorb free neutrons within the reactor core, and hence can control the rate of fissions within a reactor, stabilising energy production or shutting down the reactor completely.

c The SLCS is a back-up safety system that injects a solution of boron into the reactor to quickly bring the reactor to shut down.

d Boron is an important element as it readily absorbs neutrons.

5 **a** The ECCS is an accident response system that shuts down a nuclear reactor, even in the event of sub-system failure.

b The ECCS is activated under any accident conditions.

6 **a** The high-pressure coolant injection system is a collection of pumps used to inject coolant into the reactant vessel.

b The automatic depressurisation system consists of a range of valves that open to depressurise the reactor vessel.

c The low-pressure coolant injection system is a collection of pumps that inject coolant into the reactor vessel after depressurisation.

d Spray nozzles within the reactor pressure vessel spray water directly onto the core, reducing steam production.

e A collection of pumps and spray nozzles that spray coolant into the containment vessel.

f This system provides enough water to safely cool the reactor should the reactor systems become isolated and inoperable.

7 **a** It is the first layer of protection that surrounds the nuclear fuel.

b It is designed to surround the core; the reactor vessel provides shielding to encase radiation.

c This is the large structure, typically built of steel and concrete, that surrounds the reactor vessel.

d The core-catching mechanism is a metal 'heat sink' built under the concrete floor of the reactor to absorb heat energy from the core, should it melt through the concrete floor.

8 A standby gas treatment mechanism exists to pump air from a secondary environment so as to contain any contaminated air in the reactor environment.

1 B   **2** C   **3** D

4 It is radiation that is energetic enough to free electrons from atoms or molecules.

5 In beta-minus decay, a neutron decays into a proton, an electron and an antineutrino and assumes greater stability.

6 Nuclear fusion produces few radioactive particles, in contrast to nuclear fission, which produces myriad radioisotopes with significantly long half-lives. Nuclear fusion, although requiring significant energy to sustain, requires readily available hydrogen isotopes as reactants.

Nuclear fusion releases three to four times as much energy as nuclear fission, per unit of mass reacted.

7 The moderator in a nuclear reactor serves to slow down neutrons so that they may be more effectively captured within the nucleus of the target fuel, thereby resulting in a fission event. Control rods are used to absorb thermal neutrons, to control the rate of the nuclear chain reaction.

8 The nuclear strong force binds the protons together within the nucleus, overcoming the electrostatic repulsion of the like charges.

9 To best reflect the relative sizes of the protons, neutrons and electrons, the scale needs to be reflected. Protons and neutrons are similar in diameter at 10^{-15} m. The atom itself has a diameter of 10^{-11} m; hence, the nucleus has a diameter 10000 times smaller than that of the atom. The electron has a diameter of 1/1000th of the size of a proton.

10 The binding energy of a nucleus may be found by determining the mass defect, Δm, and then using this energy value in Einstein's mass–energy equivalence relationship, $E = \Delta mc^2$. The mass defect itself is determined by finding the difference between the rest mass of the isotope and the sum of the rest masses of the subatomic components.

11 Fusion

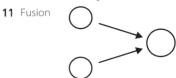

In a nuclear fusion reaction, lighter elements are fused together to form a larger element, with an accompanying release of energy.

Fission

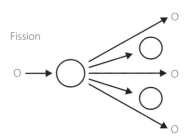

In a nuclear fission reaction, a neutron is absorbed by the nucleus of a heavy element, splitting the nucleus into two or more fission fragments as well as releasing neutrons and energy.

12 Uncontrolled nuclear fission reactions require a critical mass of the nuclear fuel, uranium, to sustain the reaction, as well as an initial source of slow thermal neutrons that can be captured by the fuel.

13 Nuclear fusion reactions release three to four times more energy that nuclear fission reactions per unit mass, but they occur under conditions of extreme pressure and temperature, and hence are very difficult to achieve.

14 For the safety of the general public, nuclear power stations would ideally be located in geographically stable areas away from the coastline. Additionally, somewhat in contradiction, they would be best positioned near large populations to reduce loss of power in energy distribution.

CHAPTER 9 REVISION

9.1 CURRENT, POTENTIAL DIFFERENCE AND ENERGY FLOW KEY TERMS

1 The current flowing through an *electric circuit* requires a *potential difference* placed across a *conductor*, typically a material with a *metal lattice* with free *electrons* able to flow. This *current* is generally termed '*conventional current*', considering the convention of flowing positive charge, as opposed to the actual flow of electrons. Such a current may alternate direction, or flow in a single direction, termed *direct current*.

2 Students answers will vary but should include the following.

Electron	Has a negative charge
Proton	Has a positive charge
Neutron	Has no charge

Static electricity is the result of the relocation of electric charges from one object to another.

In accordance with the law of conservation of charge, the net charge within a closed system remains constant, although it may be locally positive or locally negative.

9.2 CHARGE CARRIERS AND STATIC ELECTRICITY

1 The rubbing of two surfaces together allows the transfer of charge from one object to another, dependent upon their affinity for electrons.

2 Rubbing two perspex rods with a woollen cloth will allow both rods to become charged similarly. Placing one charged rod onto a rotating platform will allow the second charged rod to be brought near it, repelling it and allowing it to confirm that like charges repel.

3

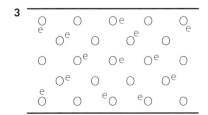

9780170412551

4

SUBATOMIC PARTICLE	CHARGE	MASS
Electrons	-1 or -1.6×10^{-19} C	9.11×10^{-31} kg
Protons	$+1$ or $+1.6 \times 10^{-19}$ C	1.673×10^{-27} kg
Neutrons	Nil	1.675×10^{-27} kg

5 Static electricity refers to charge that is 'at rest' being localised on an insulating material, such as perspex. Current electricity requires a conductor and refers to electric charge that is able to flow.

6 Examples:

1 Static electricity discharges from a vehicle door

2 Rubbing a balloon on your head will attract lengths of hair

3 Removing a woolly jumper in winter creates static discharges

9.3 CURRENT ELECTRICITY AND ELECTRICAL DEVICES

1 Electric *circuits* provide a pathway for *current* to flow. Any *gap* in the circuit will prevent this from happening. Devices such as *globes* convert electrical *energy* into other forms of energy such as *heat* or *light*.

Electrical energy can be produced using *chemical reactions* or *generators*. The rate at which electrical energy is being used is termed '*power*', which is measured in *watts*, named after a famous British scientist.

2

TABLE 9.3 Characteristics of electrical devices

DEVICE	SYMBOL	FUNCTION OF DEVICE
Battery		A power source based on chemical energy
Voltmeter	$-\text{(V)}-$	A device for measuring voltage across a device
Switch	$-\circ\!\!\!\diagup\!\circ-$	A device for controlling a circuit (off or on)
Globe		A device that converts electrical energy into light and often heat
Resistor		The opposition to electric flow due to the characteristics of the material
Ammeter	$-\text{(A)}-$	A device for measuring the current through a circuit

3 Residual current devices (RCDs) monitor the flow of current in a circuit and, if a loss of current is detected, the circuit is shut off, preventing a fatal electric shock.

Insulated electrical cords and insulated metal pins offer further protection against accidental electric shock.

Fuses are placed in household circuits to purposely burn out should an excess current flow in the circuit.

Circuit breakers are magnetic devices that are current sensitive, tripping when an overload of current is experienced.

4 Energy is the capacity to do work. It is measured in joules, J, and may be found in a range of forms including kinetic (moving) and potential (electrical, chemical and gravitational potential). Power, measured in watts, W, is a quantity of work done in a given period of time.

5 a Electricity consumption is typically charged by the kWh. A typical household of four people will use approximately 9000 kWh per annum. Queensland energy is currently charged at the rate of 22.8 c per kWh.

b $1 \text{kWh} = 1000 \text{J s}^{-1} \times 3600 \text{s} = 3\,600\,000$ J.

6 For current electricity to flow, it requires a conducting path, typically metal, such as copper.

7 A switch is a device placed within a circuit to control the flow of current, allowing the path to be broken (off) or reconnected (on).

8 Electrical energy can be converted into other forms of energy such as heat or light.

The battery provides a source of energy for an electric circuit.

The rate at which electricity flows around a circuit is called the electric current.

9

ELECTRICAL APPLIANCE	FORM OF ENERGY TRANSFORMATION
Hair dryer	Electrical → kinetic and radiant heat
Toaster	Electrical → radiant heat
Fan	Electrical → kinetic
Television	Electrical → radiant light and sound
Phone charger	Electrical → chemical potential
Oven	Electrical → radiant heat and light
Heater	Electrical → radiant heat energy
Fan	Electrical → kinetic energy
Television	Electrical → radiant light and heat
Stereo	Electrical → radiant sound (kinetic energy of air particles)
DVD player	Electrical → radiant light, kinetic energy
Refrigerator	Electrical → kinetic energy (pumps and circulation of refrigerant), radiant heat and sound
Freezer	Electrical → kinetic energy (pumps and circulation of refrigerant), radiant heat and sound
Air conditioner	Electrical → kinetic energy (pumps and circulation of refrigerant), radiant heat and sound

10 Electricity and the storage of electrical energy has improved efficiency in all manner of electrical devices, including increasing portability and flexibility in communication devices.

Communication, entertainment, and commercial business, indeed, world economies are increasingly dependent upon the internet and electrical circuits for their effective operation.

■ 9.4 ELECTRICAL CIRCUIT SYMBOLS

Draw the circuit symbols for the following electrical devices.

TABLE 9.4.1 Electrical circuit symbols

ELECTRICAL DEVICE	CIRCUIT SYMBOL
Switch (open)	
Fixed resistor	
Lamp	
LED	
Battery of cells	
Ammeter	
Voltmeter	
AC supply	
Cell	
Earth	

■ 9.5 SERIES AND PARALLEL CIRCUITS

1 a Ohm's law

b $V = I \times R$, or $R = V/I$

c R is inversely proportional to I; that is, as the resistance within a circuit increases, the current flowing through the circuit decreases.

2 A series circuit contains devices in a single path, that is, the current does not split; hence, all devices within the circuit experience the same current.

A parallel circuit provides multiple paths for the flow of current, allowing it to split, not necessarily evenly, between independent circuit loops.

3
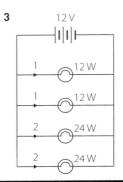

$P = V \times I$

$24\,\text{W} = 12\,\text{V} \times I$

Therefore $I = 2\,\text{A}$

$P = V \times I$

$12\,\text{watt} = 12\,\text{volt} \times I$

Therefore $I = 1\,\text{A}$

4 Given that LEDs are 90% efficient, and incandescent light globes are 5% efficient, LEDs are 18 times more effective at converting electrical energy to radiant light energy; hence, LEDs are potentially 18 times brighter than incandescent light globes.

5 Car headlights are of paramount importance for vehicle safety; hence, all lights need to be able to operate independently, on parallel circuits.

■ 9.6 DC AND AC CIRCUITS

1 Direct current flows in a single direction throughout a conducting circuit and is typically provided by portable, battery sources. Alternating current, as the name indicates, alternates the direction of flow. In Australian mains voltage supply, the current oscillates at a frequency of 50 Hz.

2 DC circuit symbol   AC circuit symbol

3 Mobile phones, portable speakers, watches

4 $V = 240\,\text{V rms}, f = 50\,\text{Hz}$

■ 9.7 ELECTRICAL CIRCUIT ANALYSIS

1 a $I = Q/t$, hence 30 coulomb/60 seconds = 0.5 A

b X: Voltmeter (voltage, volts)

Y: Ammeter (current through total circuit, amperes)

Z: Ammeter (current through resistor, amperes)

c $V = 3.0\,\text{V}, I = 0.5\,\text{A}$, hence $R = V/I = 6\,\Omega$

d X: 3.0 V

Y: 1.5 A

Z: 0.5 A

2 a $P = V \times I = 240\,\text{V} \times 6\,\text{A} = 1440\,\text{watt}$

b $R = V/I = 240\,\text{V}/6\,\text{A} = 40\,\Omega$

c Typical use of the kettle requires 6 A of current. (Household circuits are set to a maximum of 10 A.)

d i 6 A

ii 0 A

e $9.18

3 a Thinner wires offer greater resistance; hence, they heat up more readily and give off light.

b Tungsten is a dense, hard conducting metal with a high resistance and very high melting point.

c To reduce the loss of charge, current and heat energy from the filament to the external environment.

d Electrical energy is transformed into radiant light and radiant heat.

■ 9.8 ELECTRICAL APPLIANCES

1 A

2 D

3 Washing machine (A)

4 Hair curling iron (E)

5 B

6 C

9 C

CHAPTER 9 EVALUATION

1 a True

b True

c False. A 15-watt light globe uses 15 joules of energy every second.

d False. When a vehicle doubles its speed, its kinetic energy is quadrupled.

e True

f True. Although energy is conserved, in reality, energy is 'lost' to the surroundings in any transformation.

g True

h True

i True

2 B

3 B

4 $KE_{Truck} = 312\,500\,J$

$KE_{Car} = 900\,000\,J$

The car has greater kinetic energy.

5

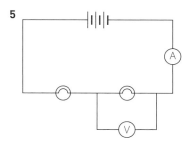

6 Long-distance space travel will take a significant time period; hence, a small force applied over long period of time, with no air resistance, could accelerate a craft to very high velocities.

7 Work is synonymous with energy – they are both measured in joules. Energy is the capacity to do work; to do work requires the transformation of energy. Power is the rate at which energy is transformed, measured in joules per second, or watts.

8 a $P = V \times I = 2.1$ watt

b $R = V/I = 4.29\,\Omega$

9 5 seconds

10 Wires of greater thickness have larger cross-sectional area, and hence offer less resistance to the flow of charge.

11 $70.13

CHAPTER 10 REVISION

■ 10.1 REVISING RESISTANCE

1 **Ohm's law** describes the relationship between voltage and current through a resistor. When the voltage across and current through a device is proportional, the device is termed **ohmic**; if it is not proportional, then the device is termed **non-ohmic**.

A graph of voltage versus current for a device demonstrates if it is ohmic (linear graph) or non-ohmic (non-linear).

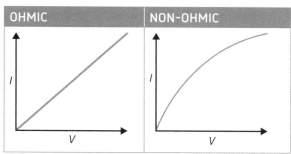

2 Student responses should include the following.

The resistivity of a conductor varies for each specific material, and determines its classification as a conductor, semiconductor or insulator. Conductors allow the flow of charge, insulators do not allow the flow of charge and semiconductors conduct electricity better than insulators, but not as well as metal conductors. Ohmic devices are conductors that follow Ohm's law, $R = V/I$.

■ 10.2 RESISTANCE

1

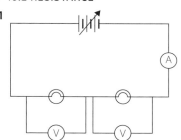

2 Globe A – there is a linear relationship between current and voltage.

3 The resistance of conductors is dependent on temperature, and hence an increase in temperature may increase the resistance, decreasing the current, and result in a non-linear graph.

4 a $2.5\,\Omega$

b As the potential difference increases, the current increases at a greater rate, and hence the resistance decreases.

5 The experiment requires electric current to be run through conductors, and hence there is a risk of electric shock. Additionally, the conductors may become very hot, so there is also a risk of burning.

6 LEDs have very little resistance, and so draw very little current. A 240 V supply across a device with little resistance would require too large a current to run through a household circuit.

■ 10.3 CONDUCTORS AND INSULATORS

1 Electrons are able to move freely between atoms in the lattice structure.

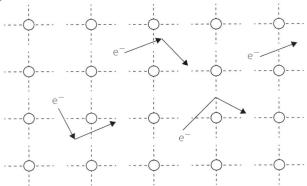

2

INSULATORS	CONDUCTORS	SEMICONDUCTORS
Plastic	Copper	Silicon
Rubber	Gold	Gallium arsenide
Wood	Silver	Aluminium gallium indium phosphide

3 Metals have a lattice structure in which the outer electrons are loosely bound and are able to move freely.

■ 10.4 RESISTANCE AND RESISTIVITY

1 Ohm's law is the proportional relationship between the current through and the voltage applied across a conductor, represented as $R = V/I$.

2 Cross sectional area, length and resistivity

3 $R = \rho \times \ell/A$.

4 An increase in the temperature of a conductor increases its resistance, reducing the current flow through the conductor.

5 The resistor is ohmic.

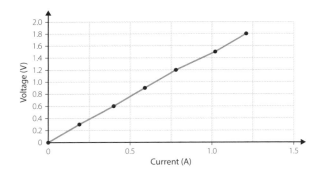

■ 10.5 SYSTEMATIC AND RANDOM ERRORS

1 Systematic experimental errors could include calibration error, offset or zero setting error, scale factor error, and parallax error.

2 Random experimental errors could include atmospheric inconsistencies (e.g. wind resistance), instrument reading fluctuations, and values lying within the scale increments.

3 Systematic errors are often termed one-sided errors, while random errors are often termed two-sided errors as they vary in a random manner, on either side of the measured value.

4 Systematic errors are often termed one-sided errors since they affect the measurements on one side of the expected value, either consistently greater than (or less than) the true value.

■ 10.6 THE INVENTION OF THE TRANSISTOR

1 Valves, or thermionic devices, are made from glass bulbs, making them extremely fragile.

2 Solid state devices were developed to make more reliable and portable transistors.

3 Solid state devices are more durable and have a longer life span than valve-based devices. They also operate at a greater rate and capacity, are more efficient, and are more economical to produce.

4 Solid state devices, such as transistors, have enabled personal, mobile communication and significantly increased computing capacity in all manner of devices.

CHAPTER 10 EVALUATION

1 B **2** D **3** A

4 a

6 V

$R_1 = 15\ \Omega$ $R_2 = 9\ \Omega$

b $R_T = 24\ \Omega$, $I_T = 0.25\ \text{A}$

c $V_1 = 3.75\ \text{V}$, $V_2 = 2.25\ \text{V}$

d

6 V

$R_T = 24\ \Omega$

5 Electrical engineer, software developer, sound engineer, electrician

6 $R = \rho \times \ell/A$

7 An additional resistor placed in parallel increases the cross-sectional area of the pathway throughout which electrons can flow, thereby increasing the current.

8 a

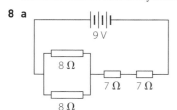

b $R_T = 18\,\Omega$, $V_T = 9\,V$, therefore $I_T = 0.5vA$

c $V_{8\Omega} = I_{8\Omega} \times R_{8\Omega}$

$V_{8\Omega} = 0.25\,A \times 8\,\Omega$

$V_{8\Omega} = 2\,V$

$V_{7\Omega} = I_{7\Omega} \times R_{7\Omega}$

$V_{7\Omega} = 0.5\,A \times 7\,\Omega$

$V_{7\Omega} = 3.5\,V$

d $P = V \times I$

$P = 9\,V \times 0.5\,A$

$P = 4.5\,W$

9

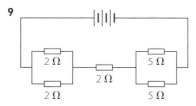

CHAPTER 11 REVISION

11.1 CIRCUIT ANALYSIS AND DESIGN KEY TERMS

1 An electric circuit is a conducting path that allows a current of charges to flow due to a *potential difference*. Circuits may include components in series, in parallel or, in the case of a *combination circuit*, with components arranged in both. Circuits may be analysed with respect to their *current* and voltage values using *Kirchhoff's current law* and *Kirchhoff's voltage law*.

2 Parallel circuits provide multiple paths for electric current, as opposed to series circuits, which provide a single pathway.

Combination circuits contain aspects that are in both series and parallel arrangements.

An emf, or electromotive force, provides the potential energy for electric charges to move, forming a current throughout the circuit.

11.2 DRAWING ELECTRIC CIRCUITS

1

2

3 A series circuit provides a single pathway for current to flow through. A parallel circuit provides multiple pathways throughout which charges may flow.

4

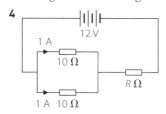

b Current = charge/time

$I = q/t$

$2 = 2.8 \times 10^5/t$

$t = 1.4 \times 10^5$ seconds

$t = 38.89$ hours

11.3 SAFETY DEVICES IN CIRCUITS

1 The ERAC aims to provide uniform regulations for electrical activities across government, industry and the public with specific attention to safety, supply, installations and inspections, quality, supply and efficiency.

2 The colour change in three-pin power point connections in Australia has been made to align with the International Electrotechnical Commission (IEC) requirements in an aim to bring all coloured wiring connections into line worldwide.

3 The earth wire provides a conductive path to the ground, allowing excessive current to flow along, should there be a fault in an electrical appliance.

4 The current required for lighting circuits is considerably less than that required for a stove circuit.

5 Using larger diameter wiring reduces the resistance, increasing the current in a household circuit. This has implications for the total current throughout the circuit, which may then exceed the 10 ampere maximum.

11.4 LIGHT-EMITTING DIODES

TABLE 11.4.1 Comparison of light sources

LIGHT SOURCE	OPERATION	TYPICAL APPLICATION	APPROXIMATE EFFICIENCY
Light-emitting diodes	High tech	Household lighting; television back lighting; traffic lights	12%
Arc lamps	Low tech	Street and factory lighting	7%
Sodium lamps	Low tech	Street and factory lighting	12%
Incandescent lamps	Low tech	Household lighting	2%
Neon/argon lamps	Low tech	Commercial signage; airport runway lighting	55%
Halogen lamps	High tech	Downlights; car headlights	3%
Fluorescent lamps	Low tech	Household lighting	9%
Metal halide	Low tech	Stadium and sports field lighting	12%
Lasers	High tech	DVD players; surgical cutting and cauterising	25%

11.5 BATTERY STORAGE SYSTEMS FOR HOUSEHOLDS

1 12 kWh

2 20

3 8.33 A

4 2000 watts × 6 hours/240 volts = 50 amp-hours

11.6 ELECTRONIC DEVICES

1 Student responses will vary.

2 Future developments in electronics will assist in making communication devices smaller, more efficient, more portable and available within the myriad of other devices, even wearable technology.

3 Varied responses available, including the miniaturisation of electronic devices enabling personal computing to be commonplace, and supporting globalisation of industry, commerce, politics, scientific communities and information gathering and distribution.

CHAPTER 11 EVALUATION

1 C   2 B   3 D

4 $R_T = R_1 + R_2 + \ldots + R_n$

5 $\dfrac{1}{R_T} = \dfrac{1}{R_1} + \dfrac{1}{R_2} + \ldots + \dfrac{1}{R_n}$

6 The sum of electrical current flowing into a junction is equal to the sum of current flowing out of the junction.

7 The sum of the potential rises within a closed loop is equal to the sum of the potential drops within the loop.

8 2.4 A

9 a 5.0 A

   b 48 Ω

   c $P = I \times I \times R$

   $P = 1200\,\text{W}$

   d $E = P \times t = 144\,000\,\text{J}$

10 a 1.67 A

   b 144 Ω

   c 400 W

   d 72 000 J

CHAPTER 12 REVISION

12.1 SCALAR AND VECTOR QUANTITIES

1 a Temperature and speed are scalars.

   b They require one scale (no direction).

2 It has magnitude (speed) and direction.

3 Scalar (straight-line motion); vector (curvilinear motion)

4 i 18 km west

   ii 48 km south

   iii 8 km west

   iv 10 km south-east

5 a 5 km south

   b 19 km west

   c 23 km east

   d 18 km south

12.2 VECTOR REPRESENTATION

1 a–b

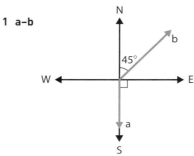

c

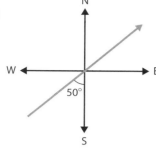

d

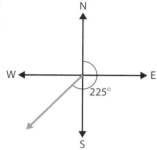

e

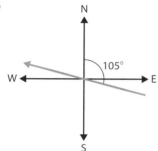

2 a

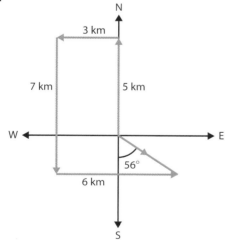

b 21 km

c **i** S(56 ± 1)°E

    ii (124 ± 1)°

3 a

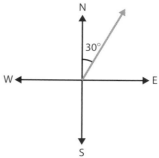

b

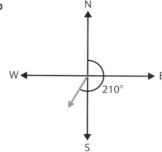

c

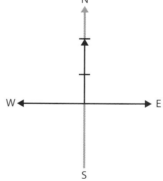

d

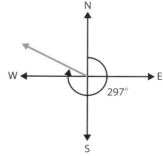

e

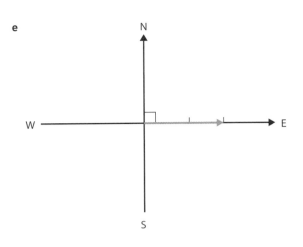

4 a

INTERSECTION	DISTANCE (m)	COMPASS ANGLE (°)
Meitner–Schmidt	1200	135
Rutherford–Meitner	900	117
Feynman–Noddack	900	27
Leavitt–Moseley	600	315
Doherty–Schmidt	1200	225

b Using east and north as positive directions, specify the easterly and northerly components relative to GirlVani's Pizza of the following intersections in Cartesia

INTERSECTION	NORTHERLY COMPONENT (m)	EASTERLY COMPONENT (m)
Lange–Feynman	0	300
Lange–Doherty	0	⁻600
Curie–Rutherford	⁻300	0
Curie–Noddack	⁺600	0
Moseley–Schmidt	⁻600	⁻300
Noddack–Doherty	⁺600	⁻600

5 a 100 m, east of Noddack–Feynman

b 721 m, N56°W

c 1270, 45°

■ **12.3 MOVEMENT ALONG A STRAIGHT LINE**

1 a ⁺11 cm; 11 cm

b ⁻34 cm; 47 cm

c ⁻6.11: 6.49 cm

d ⁺47.4 m; 65.8 m

2 a E; G; A; C; B; F; D

b ⁺80 m to ⁻100 m

c **i** ⁺40 m   **ii** ⁺80 m

   iii ⁺80 m   **iv** ⁻80 m

   v ⁻180 m   **vi** ⁻135 m

d FG = 100 m or 1 min on t-axis (v_{av} = 6 km h⁻¹ or 100 m per min). Total walk is 15 min; E is at 10 min.

1 A   **2** C   **3** C   **4** C   **5** A

6 A   **7** D   **8** D   **9** A   **10** C

11 Scalar: one scale – distance: length
Vector: more than one scale – displacement: length, direction

12 a 10 m

b ⁻4.5 m

c ⁻1.7 m

13 a

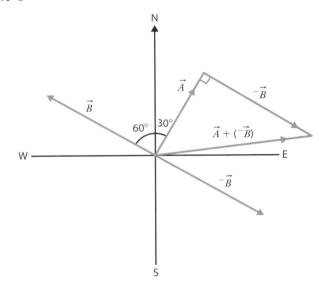

b 25 m, S

14 60 m, N(73 ± 1)°W

■ **CHAPTER 13 REVISION**

■ **13.1 DISPLACEMENT, VELOCITY AND ACCELERATION**

1 Speed is the rate of change of distance, without reference to direction. Velocity is speed with direction.

2 a ⁺0.64 m s⁻¹

b ⁻80 m s⁻¹

c ⁺37.5 m s⁻¹

d ⁺1.90 m s⁻¹

e ⁻8.5 × 10⁻² m s⁻¹

3 v_{av} = 7.1 m s⁻¹; $\vec{v}_{av}$ = 0 m s⁻¹

4 a 529 km h⁻¹

b 0 km h⁻¹

9780170412551

13.2 SPEED AND VELOCITY

1 **a** $58\,\mathrm{km\,h^{-1}}$

 b $14\,\mathrm{km\,h^{-1}}$

 c $86\,\mathrm{km\,h^{-1}}$

 d $74\,\mathrm{km\,h^{-1}}$

 e $54\,\mathrm{km\,h^{-1}}$

2 **a** $11\,\mathrm{m\,s^{-1}}$

 b $22\,\mathrm{m\,s^{-1}}$

 c $3.3\,\mathrm{m\,s^{-1}}$

 d $8.8\,\mathrm{km\,h^{-1}}$

 e $98\,\mathrm{m\,s^{-1}}$

3 $30\,\mathrm{km\,s^{-1}}$

 $= 30\,\mathrm{km\,s^{-1}} \times 3600\,\mathrm{s\,h^{-1}}$

 $= 1.1 \times 10^{5}\,\mathrm{km\,h^{-1}}$

13.3 INTERPRETING GRAPHS: LINEAR MOTION

1 **a** $20\,\mathrm{m\,s^{-1}}$

 b $20\,\mathrm{m\,s^{-1}}$

2 $40\,\mathrm{m}$

3 **a** **i** $50.4\,\mathrm{m}$

 ii $56\,\mathrm{m}$

 iii $84\,\mathrm{m}$

 iv $95.2\,\mathrm{m}$

 b **i** $1.4\,\mathrm{m\,s^{-1}}$

 ii $3.2\,\mathrm{m\,s^{-1}}$

 iii $2.8\,\mathrm{m\,s^{-1}}$

4 **a** $186\,\mathrm{m}$

 b $16\,\mathrm{m\,s^{-1}}$

13.4 ACCELERATION

1 **a** $40\,\mathrm{m}$

 b

TIME (s)	0	30	60	90	120	150	180	210	240
DISTANCE (m)	0	600	1200	1800	2400	3000	3600	4200	4800

2 **a**

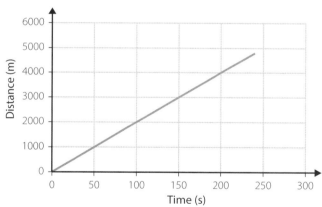

 b

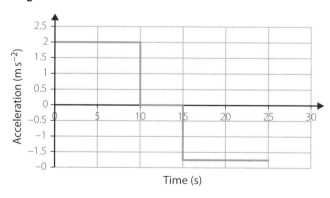

 b **i** $19\,\mathrm{m\,s^{-1}}$

 ii $18\,\mathrm{m\,s^{-1}}$

 iii $19.1\,\mathrm{m\,s^{-1}}$

3 **a** **i** $0.55\,\mathrm{m\,s^{-2}}$

 ii $9.2 \times 10^{-2}\,\mathrm{m\,s^{-2}}$

 iii $41.3\,\mathrm{m}$

 iv $2.75\,\mathrm{m\,s^{-1}}$

 b **i**

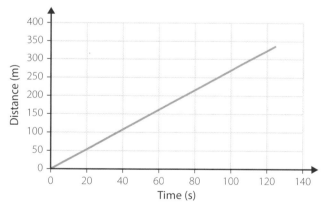

 ii

4 a area = change of speed = $15\,\text{m s}^{-1} + 8.0\,\text{m s}^{-2}(12.0\,\text{s} - 0.0\,\text{s})$
$= 111\,\text{m s}^{-1}$

b $\Delta v = a\Delta t$
$\Rightarrow \Delta v = 8.0\ \text{m s}^{-2} \times (6.2\,\text{s} - 4.0\,\text{s})$
$\Rightarrow \Delta v = 17.6\ \text{m s}^{-1}$

■ 13.5 SOLVING PROBLEMS USING ALGEBRA

1 Complete the following table.

s (m)	u (m s⁻¹)	v (m s⁻¹)	a (m s⁻²)	t (s)
35	10	5.0	1.1	4.7
83.7	15	35	6.0	3.7
320	80	0	⁻10	8.0
36	⁻57	⁺63	10	12
96	12	24.4	2.9	5.0

2 a $a = \dfrac{\Delta v}{\Delta t}$

$\Rightarrow a = \dfrac{60\ \text{m s}^{-1} - 40\ \text{m s}^{-1}}{10\ \text{s}}$

$\Rightarrow a = 2.0\ \text{m s}^{-2}$

b i $\Delta v = a\Delta t$
$\Rightarrow v - 40\ \text{m s}^{-1} = 2.0\ \text{m s}^{-2} \times 4.0\ \text{s}$
$\Rightarrow v = 40\ \text{m s}^{-1} + 8.0\ \text{m s}^{-1}$
$\Rightarrow v = 48\ \text{m s}^{-1}$

ii $v = u + at$
$\Rightarrow v = 40\ \text{m s}^{-1} + 2.0\ \text{m s}^{-2} \times 20\ \text{s}$
$\Rightarrow v = 80\ \text{m s}^{-1}$

c 500 m
3 $4.6 \times 10^{3}\,\text{m}$
4 a 4.5 s
b $9.4\,\text{m s}^{-2}$
c $5.5 \times 10^{-2}\,\text{m s}^{-2}$
5 a 64 m
b 25 m

CHAPTER 13 EVALUATION

1 A   **6** A
2 C   **7** C
3 C   **8** D
4 B   **9** B
5 C   **10** A

11 a ⁻15 km
b $14.8\,\text{m s}^{-1}$
12 a $5.0\,\text{m s}^{-1}$
b greater; steeper gradient
c ⁻$2.5\,\text{m s}^{-2}$
d 482.5 m
e $4.7\,\text{m s}^{-1}$
13 a 43.3 m
b $28.6\,\text{m s}^{-1}$
c 3.7 s
14 71.5 m

CHAPTER 14 REVISION

■ 14.1 FORCES ACTING ON AN OBJECT

1 a Contact force: objects close enough to be considered touching; Non-contact force: obvious action-at-a-distance

b Gravitational, electrostatic, magnetic

c Close encounters invoke action-at-a-distance, electrostatic force

2 Agent: object applying force; Receiver: object upon which a force is applied

3 a F(by H on J) and F(by J on H) respectively

b In the absence of any other forces: H causes J to accelerate; J causes H to accelerate.

4 a F(by C on D); F(by D on C)

b Attraction

c No, if only C and D involved; Yes, if a known charge is involved.

5 a i F(by trampoline on child)

ii F(by child on trampoline)

iii F(by gravitational field on mass of child)

b F(by trampoline on child)
$- F$(by gravitational field on mass of child) $= ma$

6 a Mass: amount of matter; Weight: force applied by a gravitational field on a mass

b Re-calibrate Earth scales to report measurements based on Moon's gravitational-field.

7 Nothing happened; Mars gravitational-field $= 2.65\,\text{N kg}^{-1}$

9780170412551

14.2 NEWTON'S THREE LAWS

1 a i $0\,\mathrm{m\,s^{-2}}$

   ii $0\,\mathrm{N}$

   iii $2000\,\mathrm{N}$

b D changes from $10\,000\,\mathrm{N}$ to $2000\,\mathrm{N}$.

2

MASS (kg)	NET FORCE ON MASS (N)	ACCELERATION ($\mathrm{m\,s^{-2}}$)
2.5	15	6.0
1.68×10^{4}	3.012	1.79×10^{-4}
0.127	0.810	6.3743
1.07×10^{-2}	852	9.138

3 Normal force is electrostatic, and weight is gravitational (Newton 3 pair must be same type).

4 Net force by winner includes F(by ground on feet of winner) $> F$(by ground on feet of loser).

5 F(by ground on athlete) $> F$(by Earth's gravitational field on athlete)

14.3 FREE-BODY DIAGRAMS

1 $10\,\mathrm{N}$

2 $4.2 \times 10^{4}\,\mathrm{N}$

3 a $2.8 \times 10^{4}\,\mathrm{N}$

  b $2.65 \times 10^{4}\,\mathrm{N}$

4 a $35.5\,\mathrm{m\,s^{-2}}$

  b $1.59\,\mathrm{kN}$

  c $2.49\,\mathrm{kN}$

5 a $1.41\,\mathrm{m\,s^{-2}}$

  b i $8.1\,\mathrm{N}$

    ii $9.6\,\mathrm{N}$

    iii $6.8\,\mathrm{N}$

14.4 CONSERVATION OF MOMENTUM

1

MASS (kg)	SPEED ($\mathrm{m\,s^{-1}}$)	MOMENTUM ($\mathrm{kg\,m\,s^{-1}}$)
4.0	6.0	24
9.32	13.7	127.8
3.43×10^{5}	6.35	2.18×10^{6}

2

MASS (kg)	TIME INTERVAL (s)	FORCE (N)	MOMENTUM CHANGE ($\mathrm{kg\,m\,s^{-1}}$)	SPEED CHANGE ($\mathrm{m\,s^{-1}}$)
8.0	4.0	6.0	24	3.0
5.88	2.6	19.6	51	8.67
1.74×10^{5}	4.72	9.77×10^{5}	4.61×10^{6}	26.5
2.4×10^{3}	10.0	1.37×10^{3}	13.7×10^{4}	5.71

3 a $3.94 \times 10^{3}\,\mathrm{N\,s}$

  b $3.94 \times 10^{3}\,\mathrm{kg\,m\,s^{-1}}$

  c $292\,\mathrm{m\,s^{-1}}$

4 Impulse (A on B) = Impulse (B on A) $\Rightarrow$ (for equal masses) velocity changes are equal and opposite

5 $0.47\,\mathrm{m\,s^{-1}}$

6 Proton: $^{-}2.6 \times 10^{4}\,\mathrm{m\,s^{-1}}$; helium: $^{+}1.54 \times 10^{4}\,\mathrm{m\,s^{-1}}$

CHAPTER 14 EVALUATION

1 C

2 C

3 C

4 D

5 C

6 B

7 C

8 A

9 B

10 C

11 B

12 B

13 D

14 C

15 $48.4\,\mathrm{m\,s^{-1}}$

16 $1.1\,\mathrm{s}$

17 a $5.0\,\mathrm{m\,s^{-2}}$

  b $3.75\,\mathrm{kN}$

  c $1500\,\mathrm{N}$

18 $4\,\mathrm{m\,s^{-1}}$

CHAPTER 15 REVISION

15.1 LAW OF CONSERVATION OF ENERGY

1

$F_{\parallel}$ (N)	X (m)	W (J)
20	10	200
24	0.76	18.04
2.21×10^{5}	0.031	6.87×10^{3}

2 $5.0\,\mathrm{km}$

3 a i $32.5\,\mathrm{kJ}$

   ii $198\,\mathrm{kJ}$

  b $183\,\mathrm{kJ}$

4

FORCE (N)	ANGLE RELATIVE TO DISPLACEMENT (°)	FORCE PARALLEL TO DISPLACEMENT (N)	DISTANCE (m)	WORK (J)
40	30	34.6	10	346
0.359	10	0.354	2.3×10^{-3}	7.9×10^{-4}
9.72×10^{3}	42.7	7.14×10^{3}	17.89	1.28×10^{5}
14	60	7.0	5.00	35
8.64	78	1.75	4.17	7.31

5 a–b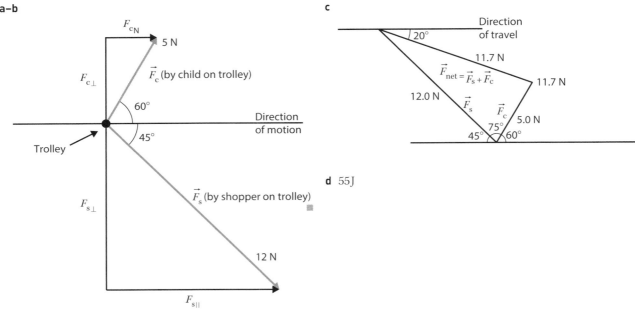

c

d 55 J

■ 15.2 ENERGY STORED IN SYSTEMS

1 $k_s : k_T = 0.94$

2

FORCE APPLIED TO SPRING (N)	FORCE APPLIED BY SPRING (N)	EXTENSION (m)	STIFFNESS (N m^{-1})	HOOKE'S LAW EQUATION	POTENTIAL ENERGY GAIN (J)
548	548	0.341	1.61×10^{3}	$F = {}^{-}1.61 \times 10^{3} x$	93.4
45	45	4.92	9.14	$F = {}^{-}9.14 x$	111
107	107	1.27	84.2	$F = {}^{-}84.2 x$	67.9
1947	1947	0.84	2.3×10^{3}	$F = {}^{-}2.3 \times 10^{3} x$	818

9780170412551

3 a $50\,\mathrm{N\,m^{-1}}$

b i $0.25\,\mathrm{J}$

ii $0.56\,\mathrm{J}$

c $2.8\,\mathrm{m\,s^{-1}}$

4

EXTENSION FORCE APPLIED (N)	LENGTH OF SPRING (cm)	SPRING EXTENSION (cm)
0.0	15.5	0.0
2.5	21.8	6.3
4.8	27.2	11.7
7.6	44.1	28.6
8.3	36.0	20.5
9.8	40.5	25

a i

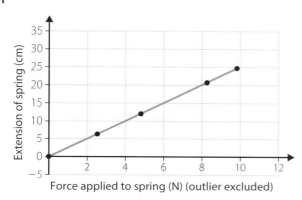

ii

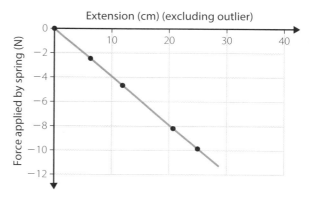

b i The extension of a spring causes the spring to apply a proportional restoring force: $F = k(^-x)$

ii x = extension of spring; F = restoring force applied by spring; k = spring constant or stiffness

iii Graph **a** (**ii**). Hooke's law relates to the force applied by the spring, which is the reaction force to the action that caused the extension (Newton 3).

c i (15 ± 0.5) cm

ii $39.2\,\mathrm{N\,m^{-1}}$

iii $1.3\,\mathrm{J}$

d No. Likely to stretch the spring beyond its elastic limit.

5

MASS (kg)	WEIGHT (N)	VERTICAL DISTANCE (m)	MAXIMUM POTENTIAL ENERGY (J)	MAXIMUM KINETIC ENERGY (J)	MAXIMUM SPEED ($\mathrm{m\,s^{-1}}$)
254	2.49×10^3	45	1.12×10^5	1.12×10^5	21
107	1050	3.50	3.67×10^3	3.67×10^3	60.6
31.5	309	6.37	1.97×10^3	1.97×10^3	7.9
1.47×10^4	1.45×10^5	6.83	9.87×10^5	9.87×10^5	9.18

■ QUESTIONS

1 a $750\,\mathrm{kg}$

b $11.3\,\mathrm{kJ}$

c $31.6\,\mathrm{m\,s^{-1}}$

2 a i $0.41\,\mathrm{kJ}$

ii $1.10\,\mathrm{kJ}$

iii $2.81\,\mathrm{kJ}$

b i $13.1\,\mathrm{m\,s^{-1}}$

ii $21.5\,\mathrm{m\,s^{-1}}$

iii $34.3\,\mathrm{m\,s^{-1}}$

■ 15.3 ELASTIC AND INELASTIC COLLISIONS

1 a $1.2\,\mathrm{m\,s^{-1}}$; left

b Inelastic; 111 J ≠ 26 J

2 a $3.4\,\mathrm{m\,s^{-1}}$

b Elastic; 41.4 J ≠ 22.8 J

3 a $2.2 \times 10^7\,\mathrm{m\,s^{-1}}$

b Inelastic; $(2.4 \neq 1.2) \times 10^{14}$ u

4 a $3.1\,\mathrm{m\,s^{-1}}$

b Inelastic; $(8.7 \neq 4.3) \times 10^4$ J

CHAPTER 15 EVALUATION

1 B

2 C

3 B

4 C

5 D

6 C

7 A

8 D

9 A

10 B

11 4.6 kN

12 a $2.3 \times 10^2 \,\mathrm{N\,m^{-1}}$

   b 13 J

   c 3.6 cm

13 a 910 J

   b 1.8 kJ

14 a $2.9 \,\mathrm{m\,s^{-1}}$

   b Inelastic: $56.6 \,\mathrm{J} \neq 53.6 \,\mathrm{J}$

CHAPTER 16 REVISION

■ 16.1 MECHANICAL WAVES

1

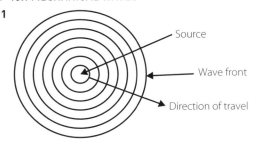

2

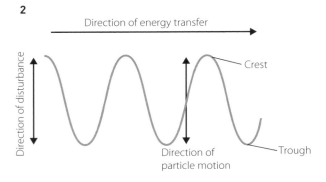

3

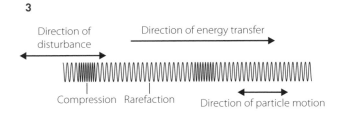

■ 16.2 WAVE FEATURES

1 a Transverse

   b Longitudinal

   c Transverse

   d Longitudinal

   e Transverse

2 a 0.10 m

   b 0.04 m

   c 0.2 s

   d 5 Hz

   e $0.2 \,\mathrm{m\,s^{-1}}$

3 $5.0 \,\mathrm{m\,s^{-1}}$

4 3500 km

■ 16.3 REFLECTION

1

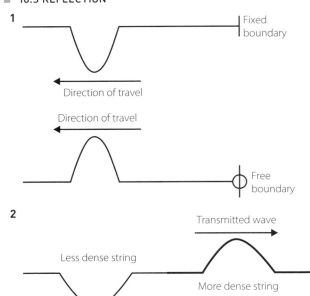

2

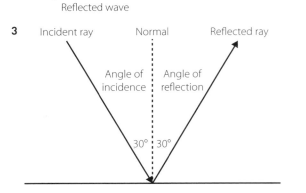

3

Incident ray   Normal   Reflected ray

Angle of incidence   Angle of reflection

30°   30°

■ 16.4 REFRACTION AND DIFFRACTION

1 The refraction of *waves* involves a change in the *direction* of the waves as they pass from one *medium* to another. *Refraction*, or bending of the path of the waves, is accompanied by changes in the *speed* and *wavelength* of the waves. This is due to the new medium having a different *elastic* property and/or *mass density* that affects the rate of *transmission* of the wave energy.

2

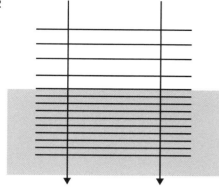

3

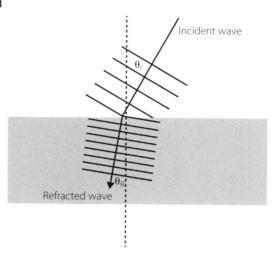

4 The concept of *diffraction* explains that when waves encounter an *obstacle* or a *gap* in a boundary, they will *bend* around it and move into the region *behind* it to some degree.

The *amount* of bending that occurs is *dependent* upon the *wavelength* of the *incident* wave and the size of the obstacle it encounters. The amount of spread is *greatest* when the wavelength is greater than the obstacle or gap width.

5 a

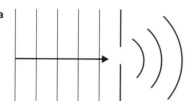

b

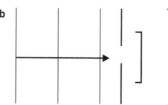

■ **16.5 THE PRINCIPLE OF SUPERPOSITION**

1 a   i Out of phase

  ii ▬▬▬▬▬▬▬▬▬▬▬▬▬▬▬▬▬▬▬▬▬▬▬▬▬▬▬▬▬▬▬

  iii Destructive interference

 b i In-phase

  ii

  iii Constructive interference

2

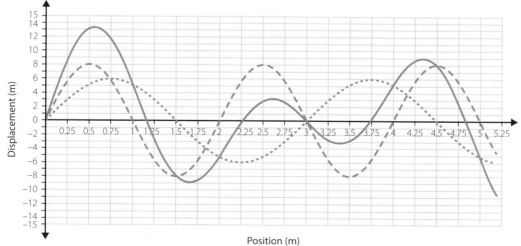

3

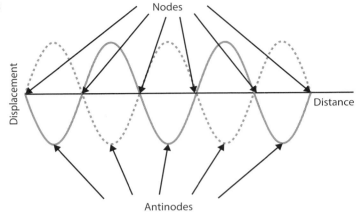

Nodes: The displacement of the particles will always be zero.

Antinodes: The displacement of the particles will oscillate between the maximum amplitude and the minimum amplitude.

CHAPTER 16 EVALUATION

1 A

2 B

3 D

4 A

5 C

6 B

7 A medium

8 In phase

9 Nodes

10

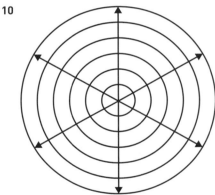

11 Responses may vary but should include mention of the following:

- Energy transfer
- Both have particles that oscillate
- Longitudinal waves oscillate in the same direction as wave trave
- Transverse waves oscillate perpendicular to direction of wave travel

12 a

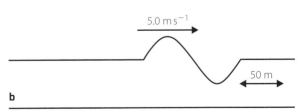

b

13 a $T = 2.500 \times 10^{-3}\,\text{s}$

　　b $v = 343.0\,\text{m s}^{-1}$

14 a $A = 0.1\,\text{m}$

　　b $T = 0.2\,\text{s}$

　　c $f = 5\,\text{Hz}$

　　d $\lambda = 1.5\,\text{m}$

　　e $v = 7.5\,\text{m s}^{-1}$

15 Responses may vary but should include mention of the following:

　　a Continuous waves

　　b Reflection at hard boundaries

　　c The principle of superposition

　　d Constructive and destructive interference

　　e Nodes and antinodes

16 Responses may vary but should include mention of the following:

- Emission of waves (sound or EM)
- Reflection at a surface
- Receiving of signal after some time
- Relationship between constant velocity and time of travel

17 Responses may vary but should include mention of the following:

- Different velocities of S and P waves
- Different velocities in different media
- Different refraction depending on media
- Different degrees of diffraction depending on size of obstacles

17.1 RESONANCE

1 a Natural frequency – the vibration frequency that occurs when an object is displaced from its equilibrium position then left to vibrate by itself

b Forced vibration – the vibration that occurs in an object when it is forced to vibrate by another vibrating object

c Resonance – when an object is induced to vibrate at its natural frequency by the vibration of another object that is also vibrating at that natural frequency

d Driving frequency – the vibration of an object that causes a second object to undergo resonance

2 · *Resonance* will only occur when the driving *frequency* matches the *natural* frequency.

· The *amplitude* of the vibration of the *resonating* object will increase dramatically.

· When an object is resonating, *energy* is being transferred with maximum *efficiency* from the driving oscillator to the receiving oscillator.

17.3 AIR COLUMNS

1

Air column	Displacement	Pressure	Frequency
Open air column			155 Hz
Closed air column			116 Hz

2 a $f_2 = 230\,Hz, f_3 = 345\,Hz, f_4 = 460\,Hz$

b $v = 345\,m\,s^{-1}$

c $\lambda_2 = 1.5\,m$

d Yes, it will form with $n = 8$

3 a $\lambda_1 = 1.3\,m$

b $f_1 = 266\,Hz$

c % diff = 20.3%

d Responses may vary but should include the following:

· Shape of bottle

· Temperature of air

· Size of embouchure

1 B

2 B

3 A

4 C

5 B

6 Resonant frequency

7 Node

8 2.6 m, 1.3 m, 0.87 m

9 a 2.4 m

b 96 Hz

c 0.30 m

17.2 VIBRATING STRINGS

1

Vibration mode	Diagram	Wavelength
1st harmonic		2.4 m
2nd harmonic		1.2 m
3rd harmonic		0.8 m

2 $f_1 = 256\,Hz$

$f_2 = 512\,Hz$

$f_3 = 768\,Hz$

$f_4 = 1020\,Hz$

3 a $\lambda_1 = 0.90\,m$

b $f_3 = 2.2\,Hz$

c Distance = 15 cm

10 a 280 Hz, 420 Hz, 560 Hz

b $504\,m\,s^{-1}$

c 1.2 m

d Yes, it will form when n = 12

11 a 10.4 m, 3.47 m, 2.08 m

b 0.743 m

c 33.0 Hz

12 $350\,m\,s^{-1}$

13 Responses may vary but should include discussion of the following:

· Driving frequency of earthquake

· Resonant frequency of building

· Efficiency of energy transfer

· Resonance

14 Responses may vary but should include discussion of the following:

· Different lengths

· Different harmonic frequencies

· Quality of tone

· Shape of instruments

■ 18.1 MODELS OF LIGHT

1 In some experiments, *light* seems to travel as a *wave*, but to interact with *matter* as a *particle*. These *experiments* cannot be explained without the use of *both* the wave and particle *models* together. Scientists call this need for these two apparently quite different models the wave–particle *duality*.

2 a The ray model

   b The wave model

   c The particle model

   d The ray model

   e The particle model

   f The wave model

3 a

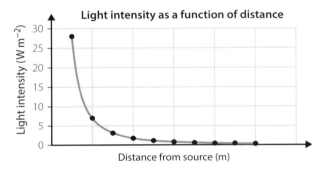

Light intensity as a function of distance

b $\text{Intensity} = \dfrac{28}{(\text{distance})^2}$

4 a

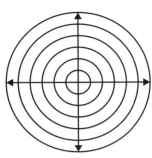

b

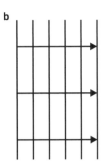

c The wavefronts in part **a** are becoming more parallel as the distance from the source increases.

■ 18.2 POLARISATION AND THE TRANSVERSE MODEL

  a Will allow light waves to pass through:

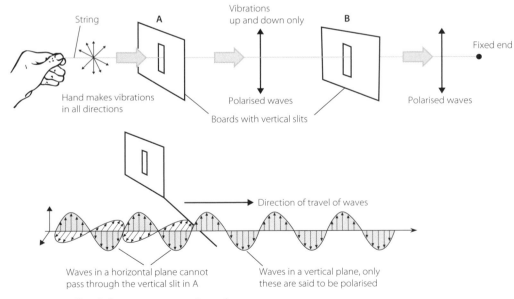

String | A | Vibrations up and down only | B | Fixed end

Hand makes vibrations in all directions

Polarised waves

Boards with vertical slits

Polarised waves

Direction of travel of waves

Waves in a horizontal plane cannot pass through the vertical slit in A

Waves in a vertical plane, only these are said to be polarised

  b Does not allow light waves to pass through:

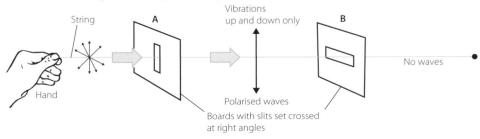

String | A | Vibrations up and down only | B | No waves

Hand

Polarised waves

Boards with slits set crossed at right angles

9780170412551

18.3 REFLECTION OF LIGHT

1 a

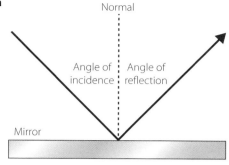

b

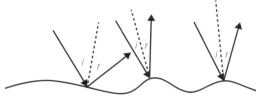

2

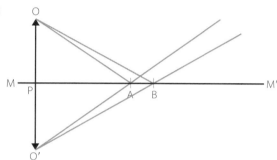

a The image will be virtual since PO′ lies inside the mirror and cannot be projected onto a screen.

b Since OP = PO′, $M = \dfrac{\text{PO}'}{\text{OP}} = 1$

18.4 SNELL'S LAW AND THE REFRACTION OF LIGHT

1 When a ray of *light* travels from one transparent *medium* into another, it changes direction. This phenomenon is called *refraction*. The amount of refraction is mainly related to differences in the *electrical* properties of each medium. The electromagnetic wave changes *speed* depending on how well the electromagnetic wave is *permitted* to move through the medium.

2 a

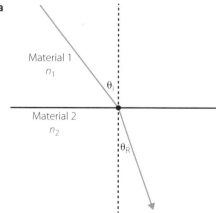

b

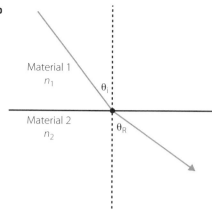

3 a 390 nm

b 32°

c

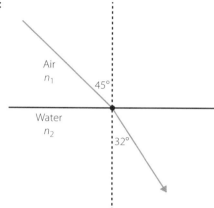

d $3.0 \times 10^8\,\text{m s}^{-1}$

4

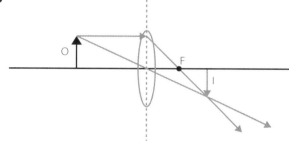

a 12 cm on the opposite side of the lens

b Real, inverted

c 3.0 cm

d $M = 1$

5

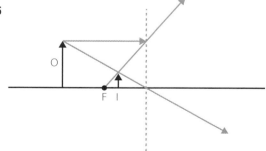

a 4.0 cm on the same side of the lens

b Virtual, upright

c 1.0 cm

d $M = 0.33$

18.5 DIFFRACTION

1

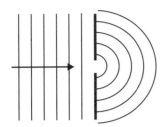

2

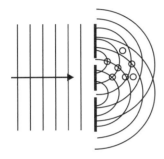

3

DISTANCE FROM SLIT 1 (nm)	DISTANCE FROM SLIT 2 (nm)	CONSTRUCTIVE OR DESTRUCTIVE
500	1000	Constructive
500	1250	Destructive
500	1300	Neither
1250	1750	Constructive
2560	3060	Constructive
2560	2810	Destructive
2560	3560	Constructive

1 B

2 C

3 D

4 A

5 C

6 D

7 Wave–particle duality

8 Photon

9 Incident wave

10 $37.5\,\mathrm{W\,m^{-2}}$

11 Angle of incidence = angle of reflection

12 Speed, angle to the normal, wavelength; $109.38\,\mathrm{W\,m^{-2}}$

13 a 592 nm

b 41.7°

c $2.26 \times 10^8\,\mathrm{m\,s^{-1}}$

d 48.8°

14 a 30 cm

b Virtual, upright

c 10 cm

d 2

15 Responses may vary but should include mention of the following:
   - Placing screen a known distance from the double slits
   - Measuring the distance between bright spots
   - Relating this distance to differences in path length
   - Using path length difference for constructive interference, pd $= n\lambda$, to calculate the wavelength

16 Responses may vary but should include mention of the following:
   - Core
   - Cladding
   - Relative refractive index <1
   - Critical angle
   - Total internal reflection

PRACTICE EXAMINATION ANSWERS

PHYSICS UNITS 1 & 2

▒ MULTIPLE-CHOICE QUESTIONS

1 A
2 D
3 D
4 B
5 C
6 C
7 D
8 C
9 C
10 C
11 C
12 B
13 D
14 A
15 B
16 A
17 B
18 A
19 B
20 D

▒ SHORT-RESPONSE QUESTIONS

1 296 K
2 5 966 L
3 Nuclear fusion
4 Sievert
5 5.0 V; Kirchhoff's voltage law.
6 85.56
7 60 days
8 The energy required to separate completely all the nucleons in a nuclide from each other.
9 1.741×10^{-25} kg
10 1.25×10^{19}
11 $s = ut + \dfrac{1}{2}$ at; $v^2 = u^2 + 2as$
12 90 km h^{-1}; 25 m s^{-1}
13 9.07 m s^{-1}
14 250 Hz
15 $\theta_i = \theta_r$
16 5.21 m s^{-1}

17 Mass is a measurement of the amount of matter in an object, measured in kilograms. Weight is a force, determined using the mass and the acceleration due to gravity, measured in Newton.
18 0.139 m
19 2.94 m
20 0.327 m

▒ COMBINATION-RESPONSE QUESTIONS

1 42 000 kJ; 1670 kJ; 315 000 kJ; 358 670 MJ
2 153.75 kJ
3 28 716 years
4 176.4 MeV
5 25 globes
6 a

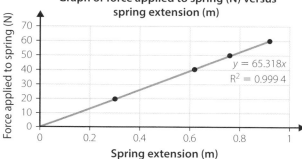

Graph of force applied to spring (N) versus spring extension (m)

$y = 65.318x$
$R^2 = 0.999\,4$

   b 65.3 N m^{-1}
   c 5.23 J
7 a 0.03 m s^{-2}
   b 5400 m + 27 000 m = 32 400 m
8 a 7350 J
   b 7350 J
   c 17.15 m s^{-1}
   d 7.35 m
9 n = 1.53
10 Inverted; 16.15 cm; 5.19 cm